ANSI C2-1990
(Revision of ANSI C2-1987)

American National Standard

National Electrical Safety Code

Secretariat

Institute of Electrical and Electronics Engineers, Inc

Approved June 26, 1989

American National Standards Institute

1990 Edition

Corrected Edition

Fifth Printing
April 1992

ISBN 1-55937-011-4

An American National Standard implies a consensus of those substantially concerned with its scope and provisions. An American National Standard is intended as a guide to aid the manufacturer, the consumer, and the general public. The existence of an American National Standard does not in any respect preclude anyone, whether he has approved the standard or not, from manufacturing, marketing, purchasing, or using products, processes, or procedures not conforming to the standard. American National Standards are subject to periodic review and users are cautioned to obtain the latest editions.

CAUTION NOTICE: This American National Standard may be revised or withdrawn at any time. The procedures of the American National Standards Institute require that action be taken to reaffirm, revise, or withdraw this standard no later than five years from the date of publication. Purchasers of American National Standards may receive current information on all standards by calling or writing the American National Standards Institute.

Abstract

This standard covers basic provisions for safeguarding of persons from hazards arising from the installation, operation, or maintenance of 1) conductors and equipment in electric supply stations, and 2) overhead and underground electric supply and communication lines. It also includes work rules for the construction, maintenance, and operation of electric supply and communication lines and equipment.

The standard is applicable to the systems and equipment operated by utilities, or similar systems and equipment, of an industrial establishment or complex under the control of qualified persons.

This standard consists of the introduction, definitions, grounding rules, list of referenced documents, and Parts 1, 2, 3, and 4 of the 1990 edition of the National Electrical Safety Code.

Key words: communication industry safety; construction of communication lines; construction of electric supply lines; electric supply stations, electric utility stations; electrical safety; high voltage safety; operation of communication systems; operation of electric supply systems; power station equipment; power station safety; public utility safety; safety work rules; underground communication line safety; underground electric line safety.

Foreword

(This Foreword is not a part of ANSI C2-1990, American National Standard National Electrical Safety Code.)

This publication consists of the parts of the National Electrical Safety Code (NESC) currently in effect. The former practice of designating parts by editions has become impractical. In the 1977 edition, Parts 1 and 4 were 6th editions, Part 2 was a 7th edition, Part 3, a revision of the 6th edition, Part 2, Section 29, did not cover the same subject matter as the 5th edition, Part 3 (withdrawn in 1970). In this 1987 edition, revisions were made in all parts. It is recommended that reference to the National Electrical Safety Code be made solely by the year on the published volume and desired part number. Separate copies of the individual parts are not available.

Work on the National Electrical Safety Code started in 1913 at the National Bureau of Standards, resulting in the publication of NBS Circular 49. The last complete code (the 5th edition, NBS Handbook H30) was issued in 1948, although separate portions had been available at various times starting in 1938. Part 2, Definitions, and the Grounding Rules, 6th edition, were issued as NBS Handbook H81, ANSI C2.2-1960, in November 1961, but work on other parts was not actively in process again until 1970. In 1970 the C2 committee decided to delete from future editions, Rules for the Installation and Maintenance of Electric Utilization Equipment (Part 3 of the 5th edition) now largely covered by the National Electrical Code (ANSI/NFPA 70), and the Rules for Radio Installations (Part 5 of the 5th edition). The Discussion of the National Electrical Safety Code, issued as NBS Handbook H4 (1928 edition) for the 4th edition of the NESC, and as NBS Handbook H39 for Part 2 of the Grounding Rules of the 5th edition, was not published for the 6th edition. The 1981 edition included major changes in Parts 1, 2, and 3, minor changes in Part 4, and the incorporation of the rules common to all parts into Section 1. The 1984 edition was revised to update all references and to list those references in a new Section 3. Rounded metric values, for information only, were added. Gender-related terminology was deleted. Sections 1: Introduction, 2: Definitions, 3: References, and 9: Grounding Methods, were made applicable to each of the Parts 1, 2, 3, and 4. The 1987 edition was revised extensively. Definitions were changed or added. Requirements affecting grounding methods, electric supply stations, overhead line clearances and loading, underground lines, and work rules were revised.

The 1990 edition of the National Electrical Safety Code contains several major changes from the previous edition. Some part of each of the principal rules in Section 1 has been revised so as to be more specific and less likely to be misinterpreted. Definitions for fiber optic cable, handhole, and communication lines have been changed.

In Section 9, Grounding Methods, the rules for common grounding of messengers and guys on the same supporting structure have been modified for clarification. Also, an exception has been added to the rule related to ground resistance of the neutral of multiple-grounded systems so that the requirement of four grounds per mile does not apply to the underwater portion of a line.

In Part 1, Electric Supply Stations, a note regarding fence safety clearances has been added and several rules affecting control of motors have been revised and rearranged. Also, a requirement for short-circuit protection of insulated power cables has been added.

In Part 2, Overhead Lines, a major change to the method for specifying overhead line clearances has been made. An explanation of the rationale behind this significant revision has been added as Appendix A. A new rule specifying the clearance requirements for cable defined as "fiber optic-supply" has been added. Requirements for clearances of overhead lines from grain bins have been added, as has an alternate method for determining the strength requirements for wood structures, consistent with the method for structures of other materials. Section 28, Miscellaneous Requirements, has been deleted and the rules that it contained moved to other sections so that related rules are located close to each other.

In Part 3, Underground Lines, grounding requirements for underground systems, similar to those for overhead systems, have been added. Requirements governing the use of an insulated jacket over a grounded conductor in random lay applications have also been added.

In Part 4, Work Rules, many identically worded rules, previously stated separately for supply system employees and communication system employees, have been combined in order to eliminate duplication. Where the requirements differ for the two groups of employees, the rules are stated separately. All of the tables on personnel approach and clearance distances have been updated. The rule on tagging electric supply circuits has been modified by the addition of a reference to Supervisory Control and Data

Aquisition Systems (SCADA).

The Institute of Electrical and Electronics Engineers, Inc, was designated as the administrative secretariat for C2 in January 1973, assuming the functions formerly performed by the National Bureau of Standards.

Comments on the rules and suggestions for their improvement are invited, especially from those who have experience in their practical application. In future editions every effort will be made to improve the rules, both in the adequacy of coverage and in the clarification of requirements. Comments should be addressed to:

Secretary
National Electrical Safety Code Committee
Institute of Electrical and Electronics Engineers, Inc.
445 Hoes Lane
P.O. Box 1331
Piscataway, NJ 08855-1331

A representative Committee on Interpretations has been established to prepare replies to requests for interpretation of the rules contained in the code. Requests for interpretation should state the rule in question as well as the conditions under which it is being applied. Interpretations are intended to clarify the intent of specific rules and are not intended to supply consulting information on the application of the code. Requests for interpretation should be addressed to:

Secretary—Interpretations
National Electrical Safety Code Committee
Institute of Electrical and Electronics Engineers, Inc.
445 Hoes Lane
P.O. Box 1331
Piscataway, NJ 08855-1331

If the request is suitable for processing, it will be sent to the Interpretations Committee. After consideration by the committee, which may involve many exchanges of correspondence, the inquirer will be notified of its decision. Decisions will be published from time to time in cumulative form and may be ordered.

The code as written is a voluntary standard. However, some editions and some parts of the code have been adopted, with and without changes, by some state and local jurisdictional authorities. To determine the legal status of the National Electrical Safety Code in any particular state or locality within a state, the authority having jurisdiction should be contacted.

Standards Committee Membership

At the time this standard was approved, American National Standards Committee C2 had the following membership:

A. L. Clapp, *Chairman* **F. A. Denbrock,** *Vice Chairman*
V. Condello, *Secretary*

Organization Represented	*Name*
American Insurance Service Group, Inc	E. S. Charkey
American Public Power Association	W. B. Roelle
	L. Bruneel (*Alt*)
American Public Transit Association	G. S. Pristach
Association of American Railroads	L. M. Himmel
Association of Edison Illuminating Companies	R. B. Bartlett
	W. F. Logan (*Alt*)
Bonneville Power Administration US Department of Energy	E. H. Bennett
Bureau of Reclamation, US Department of the Interior	L. Julson
Edison Electric Institute	G. W. Cock
	M. C. Mingoia (*Alt*)
	M. F. Borleis (*Alt*)
Electronic Industries Association	C. D. Hansell
Exchange Carriers Standards Association	O. C. Amrhyn
	O. J. Gusella, Jr (*Alt*)
Institute of Electrical and Electronics Engineers	F. A. Denbrock
	J. C. Engimann (*Alt*)
	V. R. Lawson (*Alt*)
International Association of Electrical Inspectors	*Vacant*
International Association of Government Labor Officials	B. O'Neill
	E. C. Lawry (*Alt*)
International Brotherhood of Electrical Workers	J. L. Dushaw
	J. M. Ozzello (*Alt*)
International Municipal Signal Association	W. Farrell
National Association of Regulatory Utility Commissioners	L. L. Smith
National Cable Television Association	J. Kearney
	R. B. James (*Alt*)

Organization Represented	*Name*
National Electrical Contractors Association	O. L. Davis
	C. J. Hart (*Alt*)
National Electrical Manufacturers Association	*Vacant*
	C. E. Burtner (*Alt*)
National Safety Council	P. Schmidt
Rural Electrification Administration US Department of Agriculture	A. W. Cain
	G. J. Bagnall (*Alt*)
Tennessee Valley Authority	R. B. Lee
	C. L. Clem (*Alt*)
Western Area Power Administration US Department of Energy	G. D. Birney
Individual Member	A. L. Clapp
Canadian Standards Association (Liaison Representative)	S. W. Guzik

SC1 Purpose, Scope, Application, Definitions, and References: Sections 1, 2, and 3

W. A. Thue, *Chairman*

A. W. Cain, *Secretary*

O. C. Amrhyn	(SC6)
F. S. Young	(*Alt*)
A. W. Cain	(SC2)
G. W. Horres, Jr	(*Alt*)
F. A. Denbrock	(SC5)
A. L. Clapp	(*Alt*)
L. Julson	(SC3)
G. R. Engmann	(*Alt*)
W. B. Roelle	(SC4)
R. H. Martineau	(*Alt*)
W. A. Thue	(SC7)
F. W. Koch	(*Alt*)
J. M. Van Name	(SC8)
J. L. Dushaw	(*Alt*)

SC2 Grounding Methods: Section 9

A. W. Cain, *Chairman*

G. W. Horres, Jr, *Secretary*

J. L. Dushaw
C. B. Gustafson
C. D. Hansell
L. Levey
J. N. Perrett, Jr
E. T. Robeson
M. M. Dixon (*Alt*)
D. W. Zipse

SC3 Electric Supply Stations: Sections 10–19

L. Julson, *Chairman*

G. R. Engmann, *Secretary*

M. J. Behling
L. A. Belfore
F. D. Burton
W. L. Johnson
A. E. Kollar
S. J. Kollman
J. M. Ozzello
D. J. Shelton
W. K. Knight, Jr (*Alt*)
K. C. Wolohon

SC4 Overhead Lines—Clearances: Section 23

W. B. Roelle, *Chairman*

R. H. Martineau, *Secretary*

O. C. Amrhyn
E. H. Bennett
W. N. Breedlove, Jr
A. L. Clapp
R. C. Gray
D. W. Hanson
D. E. Hooper
P. B. Lee
S. A. Olinick
D. G. Komassa (*Alt*)
R. G Oswald
E. W. Overstreet
J. M. Overton
J. M. Ozzello
R. H. Stevens
V. J. Warnock

SC5 Overhead Lines—Strength and Loading: Sections 24, 25, and 26

F. A. Denbrock, *Chairman*

A. L. Clapp, *Secretary*

F. B. Dewey
J. C. Engimann
F. G. Picciano (*Alt*)
W. F. Fuller
F. D. Swing (*Alt*)
J. G. Hanson
D. G. Heald
H. N. Johnson, Jr
R. A. Kravitz
R. Marsico
D. E. Hill (*Alt*)
W. A. Swanson
G. M. Wilhoite

SC6 Overhead Lines—General and Insulation: Sections 20, 21, 22, and 27

O. C. Amrhyn, *Chairman*

F. S. Young, *Secretary*

R. J. Bednarz
C. C. Bleakley
H. C. Breeding
G. W. Cock
  R. Theide (*Alt*)
D. M. Lauria
D. T. Michael
E. W. Overstreet
J. M. Ozzello
O. W. Perkins
T. A. Pinkham
E. L. Shaffer

SC7 Underground Lines: Sections 30–39

W. A. Thue, *Chairman*

F. W. Koch, *Secretary*

O. C. Amrhyn
C. E. Burtner (*Alt for NEMA*)
E. L. Bennett
  W. E. Brison (*Alt*)
C. C. Bleakley
D. E. Bouchard
R. M. Clark
L. G. Clemons
  H. B. Peel (*Alt*)
O. L. Davis
  C. J. Hart (*Alt*)
J. L. Dushaw
R. A. Fernandez
W. E. Haug
P. M. Henkels
E. W. Overstreet
G. S. Pristach
L. L. Smith

SC8 Work Rules: Sections 40–44

J. M. Van Name, *Chairman*

J. L. Dushaw, *Secretary*

C. E. Burtner (*Alt for NEMA*)
T. E. Branan
J. M. Degen
A. J. Grimard
C. W. Grose
C. J. Hart
H. J. Kientz
L. E. Meeker
  D. M. Bradley (*Alt*)
D. W. Richmond
R. D. Shores
J. L. Thomas

Contents

SECTION PAGE

Letter Symbols for Units 44

Sec. 1. Introduction to the National Electrical Safety Code 45
- 010. Purpose 45
- 011. Scope 45
- 012. General Rules 45
- 013. Application 46
  - A. New Installations and Extensions 46
  - B. Existing Installations 46
- 014. Waiver 46
  - A. Emergency Installations 46
  - B. Temporary Overhead Installations 47
- 015. Intent 47
- 016. Effective Date 48
- 017. Units of Measure 48

Sec. 2. Definitions of Special Terms 50

Sec. 3. References 63

Sec. 9. Grounding Methods for Electric Supply and Communication Facilities 67
- 90. Purpose 67
- 91. Scope 67
- 92. Point of Connection of Grounding Conductor 67
  - A. Direct Current Systems Which Are To Be Grounded 67
    - 1. 750 volts and below 67
    - 2. Over 750 volts 67
  - B. Alternating Current Systems Which Are To Be Grounded 67
    - 1. 750 volts and below 67
    - 2. Over 750 volts 68
    - 3. Separate Grounding Conductor 69
  - C. Messenger Wires and Guys 69
    - 1. Messenger Wires 69
    - 2. Guys 69

3. Common Grounding of Messengers and Guys on the Same Supporting Structure 69
D. Current in Grounding Conductor 70
E. Fences 70
93. Grounding Conductor and Means of Connection 71
A. Composition of Grounding Conductors 71
B. Connection of Grounding Conductors 72
C. Ampacity and Strength 72
1. System Grounding Conductors for Single-Grounded Systems 72
2. System Grounding Conductors for Multigrounded Alternating Current Systems 72
3. Grounding Conductors for Instrument Transformers 73
4. Grounding Conductors for Primary Surge Arresters 73
5. Grounding Conductors for Equipment, Messenger Wires, and Guys 73
6. Fences 74
7. Bonding of Equipment Frames and Enclosures 74
8. Ampacity Limit 74
9. Strength 74
D. Guarding and Protection 75
E. Underground 75
F. Common Grounding Conductor for Circuits, Metal Raceways, and Equipment 76
94. Grounding Electrodes 76
A. Existing Electrodes 77
1. Metallic Water Piping System 77
2. Local Systems 77
3. Steel Reinforcing Bars in Concrete Foundations and Footings 77
B. Made Electrodes 77
1. General 77
2. Driven Rods 78
3. Buried Wire, Strips, or Plates 78
4. Pole Butt Plates and Wire Wraps 79
5. Concentric Neutral Cable 80

6. Concrete-Encased Electrodes....80
95. Method of Connection to Electrode....81
A. Ground Connections....81
B. Point of Connection to Piping Systems....82
C. Contact Surfaces....83
96. Ground Resistance....83
A. Requirements....83
1. Supply Stations....83
2. Single-Grounded (Unigrounded or Delta) Systems....84
3. Multiple-Grounded Systems....84
97. Separation of Grounding Conductors....84
98. Number 98 not used in this edition....86
99. Grounding Methods for Telephone and Other Communication Apparatus on Circuits Exposed to Supply Lines or Lightning....86
A. Electrode....86
B. Electrode Connection....87
C. Bonding of Electrodes....87

Part 1. Rules for the Installation and Maintenance of Electric Supply Stations and Equipment

Sec. 10. Purpose and Scope of Rules....88
100. Purpose....88
101. Scope....88

Sec. 11. Protective Arrangements in Electric Supply Stations....88
110. General Requirements....88
A. Enclosure of Equipment....88
B. Rooms and Spaces....89
1. Construction....89
2. Use....89
3. Ventilation....89
4. Moisture and Weather....89
C. Electric Equipment....89
111. Illumination....90
A. Under Normal Conditions....90
B. Emergency Lighting....90
C. Fixtures....90

D. Attachment Plugs and Receptacles for General Use 90
E. Receptacles in Damp or Wet Locations 90
112. Floor, Floor Openings, Passageways, Stairs 93
A. Floors 93
B. Passageways 93
C. Railings 93
D. Stair Guards 93
E. Top Rails 93
113. Exits 93
A. Clear Exits 93
B. Double Exits 93
C. Exit Doors 93
114. Fire Extinguishing Equipment 94

Sec. 12. Installation and Maintenance of Equipment 94
120. General Requirements 94
121. Inspections 94
A. In-Service Equipment 94
B. Idle Equipment 94
C. Emergency Equipment 94
D. New Equipment 94
122. Guarding Shaft Ends, Pulleys, Belts, and Suddenly Moving Parts 95
A. Mechanical Transmission Machinery 95
B. Suddenly Moving Parts 95
123. Protective Grounding 95
A. Protective Grounding or Physical Isolation of Noncurrent-Carrying Metal Parts 95
B. Grounding Method 95
C. Provision for Grounding Equipment During Maintenance 95
124. Guarding Live Parts 96
A. Where Required 96
B. Strength of Guards 96
C. Types of Guards 96
1. Location or Physical Isolation 96
2. Shields or Enclosures 97
3. Railings 97
4. Mats 97

5. Live Parts Below Supporting Surfaces for Persons....97
6. Insulating Covering on Conductors or Parts....97

125. Working Space About Electric Equipment....106
A. Working Space (600 Volts or Less)....106
1. Clear Spaces....106
2. Access and Entrance to Working Space....106
3. Working Space....106
4. Headroom Working Space....106
5. Front Working Space....106
B. Working Space Over 600 Volts....106

126. Equipment for Work on Energized Parts....107

127. Classified Locations....107
A. Coal-Handling Areas....107
B. Flammable and Combustible Liquids....110
C. Flammable Liquid Storage Area....110
D. Loading and Unloading Facilities....110
1. Static Protection....111
2. Stray Currents....111
3. Container Filling Facilities....112
E. Gasoline Dispensing Stations....112
F. Boilers....113
G. Gaseous Hydrogen Systems for Supply Equipment....113
H. Liquid Hydrogen Systems....114
I. Sulfur....114
J. Oxygen....114
K. Liquefied Petroleum Gas (LPG)....115
L. Natural Gas (Methane)....115

128. Identification....125

129. Mobile Hydrogen Equipment....125

Sec. 13. Rotating Equipment....125

130. Speed Control and Stopping Devices....125
A. Automatic Overspeed Trip Device for Prime Movers....125
B. Manual Stopping Devices....125
C. Speed Limit for Motors....125
D. Number 130D not used in this edition....126

E. Adjustable-Speed Motors 126
F. Protection of Control Circuits....126
131. Motor Control 126
132. Number 132 not used in this edition....126
133. Short Circuit Protection....126

Sec. 14. Storage Batteries....127
140. General 127
141. Location....127
142. Ventilation 127
143. Racks 127
144. Floors in Battery Areas....127
145. Illumination for Battery Areas 127
146. Service Facilities....128
147. Number 147 not used in this edition....128

Sec. 15. Transformers and Regulators 128
150. Current-Transformer Secondary Circuits Protection When Exceeding 600 Volts 128
151. Grounding Secondary Circuits of Instrument Transformers.... 128
152. Location and Arrangement of Power Transformers and Regulators....128
A. Outdoor Installations 128
B. Indoor Installations.... 129

Sec. 16. Conductors 129
160. Electrical Protection....129
A. Overcurrent Protection Required 130
B. Grounded Conductors.... 130
C. Insulated Power Cables 130
161. Mechanical Protection 130
162. Isolation....130
163. Conductor Terminations 130
A. Insulation 130
B. Metal-Sheathed or Shielded Cable 130

Sec. 17. Circuit Breakers, Reclosers, Switches, and Fuses....131
170. Arrangement....131
171. Application....131

172. Circuit Breakers, Reclosers and Switches Containing Oil 131
173. Switches and Disconnecting Devices 132
A. Capacity 132
B. Provisions for Disconnecting 132
C. Visible Break Switch 132
174. Disconnection of Fuses 132

Sec. 18. Switchgear and Metal Enclosed Bus 132
180. Switchgear Assemblies 132
A. General Requirements for All Switchgear 132
B. Metal Enclosed Power Switchgear 133
C. Dead Front Power Switchboards 135
D. Motor Control Centers 135
E. Control Switchboards 135
181. Metal Enclosed Bus 136
A. General Requirements for All Types of Bus 136
B. Isolated Phase Bus 136

Sec. 19. Surge Arresters 136
190. General Requirements 136
191. Indoor Locations 136
192. Grounding Conductors 137
193. Installation 137

Part 2. Safety Rules for the Installation and Maintenance of Overhead Electric Supply and Communication Lines

Sec. 20. Purpose, Scope, and Application of Rules 138
200. Purpose 138
201. Scope 138
202. Application of Rules 138

Sec. 21. General Requirements 138
210. Referenced Sections 138
211. Number 211 not used in this edition 138
212. Induced Voltages 139
213. Accessibility 139
214. Inspection and Tests of Lines and Equipment 139

A. When In Service....139
1. Initial Compliance With Rules....139
2. Inspection....139
3. Tests....139
4. Record of Defects....139
5. Remedying Defects....139
B. When Out of Service....140
1. Lines Infrequently Used....140
2. Lines Temporarily Out of Service....140
3. Lines Permanently Abandoned....140
215. Grounding of Circuits, Supporting Structures, and Equipment....140
A. Methods....140
B. Circuits....140
1. Common Neutral....140
2. Other Neutrals....140
3. Other Conductors....140
4. Surge Arresters....140
5. Use of Earth as Part of Circuit....140
C. Noncurrent-Carrying Parts....141
1. General....141
2. Guys....141
3. Multiple Messengers on the Same Structure....141
216. Arrangement of Switches....142
A. Accessibility....142
B. Indicating Open or Closed Position....142
C. Locking....142
D. Uniform Position....142
217. General....142
A. Supporting Structures....142
1. Protection of Structures....142
2. Steps....143
3. Identification....143
4. Obstructions....143
5. Decorative Lighting....143
B. Unusual Conductor Supports....143
218. Tree Trimming....144
A. General....144
B. At Line Crossings, Railroad Crossings, and

Limited Access Highway Crossings....144

Sec. 22. Relations Between Various Classes of Lines and Equipment....144
220. Relative Levels....144
A. Standardization of Levels....144
B. Relative Levels: Supply and Communication Conductors....144
1. Preferred Levels....144
2. Special Construction for Supply Circuits, the Voltage of Which is 600 Volts or Less and Carrying Power Not in Excess of 5 Kilowatts....144
C. Relative Levels: Supply Lines of Different Voltage Classifications....146
1. At Crossings or Conflicts....146
2. On Structures Used Only by Supply Conductors....147
D. Identification of Overhead Conductors....147
E. Identification of Equipment on Supporting Structures....147
221. Avoidance of Conflict....148
222. Joint Use of Structures....148
223. Communications Protective Requirements....148
A. Where Required....148
B. Means of Protection....148
224. Circuits of One Class Used Exclusively in the Operation of Circuits of Another Class....149
A. Overhead Communication Circuits Used Exclusively in the Operation of Supply Circuits....149
B. Supply Circuits Used Exclusively in the Operation of Communication Circuits....150
225. Electric Railway Construction....151
A. Trolley-Contact Conductor Fastenings....151
B. High-Voltage Contact Conductors....151
C. Third Rails....151
D. Prevention of Loss of Contact at Railroad Crossings at Grade....151
E. Guards Under Bridges....153

Sec. 23. Clearances 153
230. General 153
A. Application 153
B. Measurement of Clearance and Spacing 153
C. Supply Cables 153
D. Covered Conductors 154
E. Neutral Conductors 154
F. Fiber Optic Cable 154
G. Alternating and Direct Current Circuits 155
H. Constant-Current Circuits 155
J. Maintenance of Clearances and Spacings 155
231. Clearances of Supporting Structures from Other Objects 155
A. From Fire Hydrants 155
B. From Streets, Roads, and Highways 155
C. From Railroad Tracks 156
232. Vertical Clearances of Wires, Conductors, Cables, and Equipment Above Ground, Roadway, Rail, or Water Surfaces 156
A. Application 156
B. Clearance of Wires, Conductors, Cables, and Equipment Mounted on Supporting Structures 157
1. Clearance to Wires, Conductors, and Cables 157
2. Clearance to Unguarded Rigid Live Parts of Equipment 157
3. Clearance to Equipment Cases 157
4. Street and Area Lighting 157
C. Additional Clearances for Wires, Conductors, Cables, and Unguarded Rigid Live Parts of Equipment 158
1. Voltages Exceeding 22 Kilovolts 158
D. Alternate Clearances for Voltages Exceeding 98 Kilovolts Alternating Current to Ground or 139 Kilovolts Direct Current to Ground 158
1. Sag Conditions of Line Conductors 159
2. Reference Heights 159
3. Electrical Component of Clearance 159

4. Limit 160
233. Clearances Between Wires, Conductors, and Cables Carried on Different Supporting Structures 177
A. General.... 177
1. Conductor Movement Envelope....177
2. Clearance Envelope....178
B. Horizontal Clearance 178
1. Clearance Requirements 178
2. Alternate Clearances for Voltages Exceeding 98 kV Alternating Current to Ground or 139 kV Direct Current to Ground....178
C. Vertical Clearance 179
1. Clearance Requirements 179
2. Voltages Exceeding 22 Kilovolts....179
3. Alternate Clearances for Voltage Exceeding 98 Kilovolts Alternating Current to Ground or 139 Kilovolts Direct Current to Ground....179
234. Clearance of Wires, Conductors, Cables, and Equipment from Buildings, Bridges, Rail Cars, Swimming Pools, and Other Installations....193
A. Application 193
1. Vertical and Horizontal Clearances (No Wind Displacement) 193
2. Horizontal Clearances (With Wind Displacement) 193
3. Transition Between Horizontal and Vertical Clearances 194
B. Clearances of Wires, Conductors, and Cables from Other Supporting Structures 194
C. Clearances of Wires, Conductors, Cables, and Rigid Live Parts from Buildings, Signs, Billboards, Chimneys, Radio and Television Antennas, Tanks, and Other Installations Except Bridges 195
1. Vertical and Horizontal Clearances....195
2. Guarding of Supply Conductors 196
4. Supply Conductors Attached to

Buildings or Other Installations........................196
4. Communications Conductors Attached to Buildings or Other Installations........................197
5. Ladder Space........................197
D. Clearance of Supply Wires, Conductors, Cables, and Rigid Live Parts from Bridges........................197
1. Vertical and Horizontal Clearances........................197
2. Guarding Trolley-Contact Conductors Located Under Bridges........................198
E. Clearance of Wires, Conductors, or Cables Installed Over or Near Swimming Areas........................198
1. Swimming Pools........................198
2. Beaches and Waterways Restricted to Swimming........................199
3. Waterways Subject to Water Skiing........................199
F. Clearances of Wires, Conductors, Cables, and Rigid Live Parts from Grain Bins........................199
1. Grain Bins Loaded by Permanently Installed Augers, Conveyers or Elevator Systems........................199
2. Grain Bins Loaded by Portable Augers, Conveyers or Elevators........................199
G. Additional Clearances for Voltages Exceeding 22 kV for Wires, Conductors, Cables, and Unguarded Rigid Live Parts of Equipment........................200
H. Alternate Clearances for Voltages Exceeding 98 Kilovolts Alternating Current to Ground or 139 Kilovolts Direct Current to Ground........................201
1. Sag Conditions of Line Conductors........................201
2. Reference Distances........................201
3. Electrical Component of Clearance........................201
4. Limit........................202
I Clearances of Wires, Conductors, and Cables to Rail Cars........................202
J. Clearance of Equipment Mounted on Supporting Structures........................203
1. Clearance to Unguarded Rigid Live Parts of Equipment........................203
2. Clearance to Equipment Cases........................203
235. Clearance for Wires, Conductors, or Cables

SECTION *Contents* PAGE

Carried on the Same Supporting Structure 220

A. Application of Rule 220
   1. Multiconductor Wires or Cables 220
   2. Conductors Supported by Messengers or Span Wires 220
   3. Line Conductors of Different Circuits 220

B. Horizontal Clearance Between Line Conductors 220
   1. Fixed Supports 220
   2. Suspension Insulators 222
   3. Alternate Clearances for Different Circuits Where One or Both Circuits Exceed 98 Kilovolts Alternating Current to Ground or 139 Kilovolts Direct Current to Ground 222

C. Vertical Clearance Between Line Conductors 224
   1. Basic Clearance for Conductors of Same or Different Circuits 224
   2. Additional Clearances 224
   3. Alternate Clearances for Different Circuits Where One or Both Exceed 98 Kilovolts, Alternating Current, or 139 Kilovolts Direct Current to Ground 226

D. Diagonal Clearance Between Line Wires, Conductors, and Cables Located at Different Levels on the Same Supporting Structure 227

E. Clearances in Any Direction from Line Conductors to Supports, and to Vertical or Lateral Conductors, Span or Guy Wires Attached to the Same Support 227
   1. Fixed Supports 227
   2. Suspension Insulators 227
   3. Alternate Clearances for Voltages Exceeding 98 kV Alternating Current to Ground or 139 kV Direct Current to Ground 227

F. Clearances Between Supply Circuits of Different Voltage Classifications on the Same Support Arm 229

G. Conductor Spacing: Vertical Racks 230
236. Climbing Space 246
A. Location and Dimensions 246
B. Portions of Supporting Structures in Climbing Space 246
C. Support Arm Location Relative to Climbing Space 246
D. Location of Equipment Relative to Climbing Space 246
E. Climbing Space Between Conductors 246
F. Climbing Space on Buckarm Construction 247
G. Climbing Space Past Longitudinal Runs Not on Support Arms 247
H. Climbing Space Past Vertical Conductors 248
I. Climbing Space Near Ridge-Pin Conductors 248
237. Working Space 252
A. Location of Working Spaces 252
B. Dimensions of Working Spaces 252
1. Along the Support Arm 252
2. At Right Angles to the Support Arm 252
3. Vertically 252
C. Location of Vertical and Lateral Conductors Relative to Working Spaces 252
D. Location of Buckarms Relative to Working Spaces 252
1. Standard Height of Working Space 252
2. Reduced Height of Working Space 252
E. Guarding of Energized Equipment 253
F. Working Clearances from Energized Equipment 253
238. Vertical Clearance Between Certain Communication and Supply Facilities Located on the Same Structure 254
A. Equipment 254
B. Clearances in General 254
C. Clearances for Span Wires or Brackets 254
D. Clearance of Drip Loops of Luminaire or Traffic Signal Brackets 254
239. Clearances of Vertical and Lateral Facilities

from Other Facilities and Surfaces on the Same Support 257
A. General 257
B. Location of Vertical or Lateral Conductors Relative to Climbing Spaces, Working Spaces, and Pole Steps 257
C. Conductors Not in Conduit 257
D. Mechanical Protection Near Ground 257
E. Requirements for Vertical and Lateral Supply Conductors on Supply Line Structures or Within Supply Space on Jointly Used Structures 258
1. General Clearances 258
2. Special Cases 258
F. Requirements for Vertical and Lateral Communications Conductors on Communication Line Structures or Within the Communication Space on Jointly Used Structures 259
1. Clearances from Wires 259
2. Clearances from Supporting Structure Surfaces 259
G. Requirements for Vertical Supply Conductors Passing Through Communication Space on Jointly Used Line Structures 260
1. Cables Meeting Rule 230C1 260
2. Jacketed Multiple-Conductor Cables 260
3. Cables and Conductors in Conduit or Covering 260
4. Suspended from Supply Support Arm 260
5. Supply Grounding Conductors 261
6. Clearance from Through Bolts 261
7. Multiple Conductor Cables 262
H. Requirements for Vertical Communication Conductors Passing Through Supply Space on Jointly Used Structures 262
1. Metal-Sheathed Communication Cables 262
2. Communication Conductors 262
3. Communication Grounding Conductors 263
4. Separation from Through Bolts 263

I. Operating Rods 263

Sec. 24. Grades of Construction 266
240. General 266
241. Application of Grades of Construction to Different Situations 266
A. Supply Cables 266
B. Order of Grades 266
C. At Crossings 266
1. Grade of Upper Line 266
2. Grade of Lower Line 267
3. Multiple Crossings 267
D. Conflicts 267
242. Grades of Construction for Conductors 268
A. Constant-Current Circuit Conductors 268
B. Railway Feeder and Trolley-Contact Circuit Conductors 268
C. Communication Circuit Conductors Used Exclusively in the Operation of Supply Lines 268
D. Fire-Alarm Circuit Conductors 268
E. Neutral Conductors of Supply Circuits 268
F. Surge Protection Wires 269
243. Grades of Construction for Line Supports 274
A. Structures 274
B. Crossarms and Support Arms 274
C. Pins, Armless Construction Brackets, Insulators, and Conductor Fastenings 275

Sec. 25. Loading for Grades B, C, and D 276
250. General Loading Requirements and Maps 276
A. General 276
B. Combined Ice and Wind Loading 276
C. Extreme Wind Loading 276
251. Conductor Loading 280
A. General 280
B. Loading Components 280
1. Vertical Loading Component 280
2. Horizontal Loading Component 280
3. Total Loading 281
252. Loads Upon Line Supports 281

SECTION *Contents* PAGE

A. Assumed Vertical Loading ... 281
B. Assumed Transverse Loading ... 281
1. Transverse Loading from Conductors and Messengers ... 282
2. Structure Loading ... 282
3. At Angles ... 283
4. Span Lengths ... 283
C. Assumed Longitudinal Loading ... 283
1. Change in Grade of Construction ... 283
2. Jointly Used Poles at Crossings Over Railroads, Communication Lines, or Limited Access Highways ... 284
3. Dead Ends ... 284
4. Unequal Spans and Unequal Vertical Loading ... 284
5. Stringing Loads ... 284
6. Longitudinal Capability ... 284
7. Communication Conductors on Unguyed Supports at Railroad Crossings and Limited Access Highways ... 284
D. Simultaneous Application of Loads ... 285

Sec. 26. Strength Requirements ... 285
260. Preliminary Assumptions ... 285
261. Grades B and C Construction ... 286
A. Supporting Structure ... 286
1. Metal, Prestressed, and Reinforced Concrete Structures ... 286
2. Wood Structures ... 286
3. Transverse-Strength Requirements for Structures Where Side Guying Is Required, But Can Only Be Installed at a Distance ... 291
4. Longitudinal-Strength Requirements for Sections of Higher Grade in Lines of a Lower Grade Construction ... 291
B. Strength of Foundations and Settings ... 292
C. Strength of Guys and Guy Anchors ... 293
1. Metal and Prestressed Concrete Structures ... 293

2. Wood and Reinforced Concrete Poles and Structures....293

D. Crossarms....293

1. Wood Crossarms....293
2. Concrete and Metal Crossarms....294
3. Crossarms of Other Materials....294
4. Additional Requirements....294

E. Number 261E not used in this edition....295

F. Strength of Pin Type or Similar Construction and Conductor Fastenings....295

1. Longitudinal Strength....295
2. Double Pins and Conductor Fastenings....296
3. Single Supports Used in Lieu of Double Wood Pins....297

G. Armless Construction....297

1. General....297
2. Insulating Material....297
3. Other Components....297

H. Open Supply Conductors....297

1. Sizes of Supply Conductors....297
2. Sags and Tensions....297
3. Splices, Taps, and Dead-End Fittings....298
4. Trolley-Contact Conductors....298

I. Supply Cable Messengers....298

J. Open-Wire Communication Conductors....299

K. Communication Cables....299

1. Communication Cables....299
2. Messenger....299

L. Paired Communication Conductors....299

1. Paired Conductors Supported on Messenger....299
2. Paired Conductors Not Supported on Messenger....299

262. Grade D Construction....307

A. Poles....307

1. Method A....307
2. Method B....307

B. Pole Settings....308

C. Guys....308

1. General....308

2. Side Guys 308
3. Longitudinal Guys 308
4. Strength of Guys 309
5. Where Guying Is Required But Cannot Be Installed on the Crossing Pole 309
D. Crossarms 310
1. Material and Size 310
2. Double Crossarms 310
E. Brackets and Racks 310
F. Pins 310
1. Strength 310
2. Size 310
G. Insulators 310
H. Conductors 311
1. Size 311
2. Paired Conductors Without Messengers 311
3. Sags 311
4. Splices and Taps 312
I. Messengers 312
1. Strength 312
2. Sags and Tensions 312
263. Grade N Construction 317
A. Poles 317
B. Guys 317
C. Crossarm Strength 317
D. Supply-Line Conductors 317
1. Size 317
E. Service Drops 317
1. Size of Open-Wire Service Drops 317
2. Tension of Open-Wire Service Drops 318
3. Cabled Service Drops 318
F. Trolley-Contact Conductors 318
G. Communication Conductors 318
H. Street and Area Lighting Equipment 318
264. Guying and Bracing 320
A. Where Used 320
B. Strength 320
C. Point of Attachment 320
D. Guy Fastenings 320
E. Guy Markers (Guy Guards) 320

F. Electrolysis 321
G. Anchor Rods 321

Sec. 27. Line Insulation 321
270. Application of Rule 321
271. Material and Marking 321
272. Ratio of Flashover to Puncture Voltage 322
273. Insulation Level 322
274. Factory Tests 323
275. Special Insulator Applications 323
A. Insulators for Constant-Current Circuits 323
B. Insulators for Single-Phase Circuits Directly Connected to Three-Phase Circuits 323
276. Protection Against Arcing and Other Damage 323
277. Mechanical Strength of Insulators 323
278. Aerial Cable Systems 324
A. Electrical Requirements 324
B. Mechanical Requirements 324
279. Guy and Span Insulators 324
A. Insulators 324
1. Properties of Guy Insulators 324
2. Use of Guy Insulators 325
3. Corrosion Protection 325
B. Span-Wire Insulators 325
1. Properties of Span-Wire Insulators 325
2. Use of Span-Wire Insulators 326

Sec. 28. Section Number 28 not used in this edition 326

Part 3. Safety Rules for the Installation and Maintenance of Underground Electric-Supply and Communication Lines

Sec. 30. Purpose, Scope, and Application of Rules 327
300. Purpose 327
301. Scope 327

Sec. 31. General Requirements Applying to Underground Lines 327
310. Referenced Sections 327

311. Installation and Maintenance 327
312. Accessibility 328
313. Inspection and Tests of Lines and Equipment 328
A. When in Service 328
1. Initial Compliance with Safety Rules 328
2. Inspection 328
3. Tests 328
4. Record of Defects 328
5. Remedying Defects 328
B. When Out of Service 328
1. Lines Infrequently Used 328
2. Lines Temporarily Out of Service 328
3. Lines Permanently Abandoned 329
314. Grounding of Circuits and Equipment 329
A. Methods 329
B. Conductive Parts to Be Grounded 329
C. Circuits 329
1. Neutrals 329
2. Other Conductors 329
3. Surge Arresters 329
4. Use of Earth as Part of Circuit 329
315. Communication Protective Requirements 329
A. Where Required 329
B. Means of Protection 330
316. Induced Voltage 330

Sec. 32. Underground Conduit Systems 330
320. Location 331
A. Routing 331
1. General 331
2. Natural Hazards 331
3. Highways and Streets 331
4. Bridges and Tunnels 331
5. Crossing Railroad Tracks 331
6. Submarine Crossing 332
B. Separation from Other Underground Installations 332
1. General 332
2. Separations Between Supply and Communications Conduit Systems 332

3. Sewers, Sanitary and Storm 332
4. Water Lines 333
5. Fuel Lines....333
6. Steam Lines....333
321. Excavation and Backfill 333
A. Trench 333
B. Quality of Backfill 333
322. Ducts and Joints 333
A. General.... 333
B. Installation....334
1. Restraint 334
2. Joints....334
3. Externally Coated Pipe....334
4. Building Walls....334
5. Bridges 334
6. In Vicinity of Manholes....335
323. Manholes, Handholes, and Vaults 335
A. Strength 335
B. Dimensions.... 336
C. Manhole Access 336
D. Covers.... 337
E. Vault and Utility Tunnel Access....337
F. Ladder Requirements....337
G. Drainage.... 338
H. Ventilation....338
I. Mechanical Protection.... 338
J. Identification 338

Sec. 33. Supply Cable....340
330. General 340
331. Sheaths and Jackets....340
332. Shielding 340
A. General.... 340
B. Material.... 341
333. Cable Accessories and Joints....341

Sec. 34. Cable in Underground Structures 341
340. General 341
341. Installation 341
A. General.... 341

SECTION *Contents* PAGE

B. Cable in Manholes and Vaults 342
1. Supports 342
2. Clearance 342
3. Identification 343
342. Grounding and Bonding 344
343. Fireproofing 344
344. Communication Cables Containing Special Supply Circuits 344

Sec. 35. Direct Buried Cable 345
350. General 345
351. Location and Routing 345
A. General 345
B. Natural Hazards 346
C. Other Conditions 346
1. Swimming Pools 346
2. Buildings and Other Structures 346
3. Railroad Tracks 346
4. Highways and Streets 347
5. Submarine Crossings 347
352. Separations from Other Underground Structures 347
A. Horizontal Separation 347
B. Crossings 347
C. Parallel Facilities 348
D. Thermal Protection 348
353. Installation 348
A. Trenching 348
B. Plowing 348
C. Boring 348
D. Depth of Burial 349
354. Random Separation—Additional Requirements 349
A. Supply Cables or Conductors 349
B. Communication Cables or Conductors 350
C. Supply and Communication Cables or Conductors 350
1. Voltage 350
2. Grounded Conductor: Bare or with a Jacket of Radial Resistivity of 100 Meter Ohms or Less 350
3. Grounded Conductor for Jacketed Cables

with a Jacket Resistivity Exceeding 100 Meter Ohms 351
4. Ungrounded Supply Systems 351
D. Multiple Cable Systems 352
E. Protection 352

Sec. 36. Risers 352
360. General 352
361. Installation 353
362. Pole Risers—Additional Requirements 353
363. Pad-Mounted Installations 353

Sec. 37. Supply Cable Terminations 353
370. General 353
371. Support at Terminations 354
372. Identification 354
373. Clearances in Enclosures or Vaults 354
374. Grounding 355

Sec. 38. Equipment 355
380. General 355
381. Design 355
382. Location in Underground Structures 356
383. Installation 357
384. Grounding 357
385. Identification 357

Sec. 39. Installation in Tunnels 357
390. General 357
391. Environment 358

Part 4. Rules for the Operation of Electric-Supply and Communications Lines and Equipment

Sec. 40. Purpose and Scope 359
400. Purpose 359
401. Scope 359
402. Referenced Sections 359

SECTION *Contents* PAGE

Sec. 41. Supply and Communications Systems—Rules for Employers....359
410. General Requirements....359
A. General....359
B. Emergency Procedures and First Aid Rules....360
C. Responsibility....360
411. Protective Methods and Devices....360
A. Methods....360
B. Devices and Equipment....361
C. Inspection and Testing of Protective Devices....361
D. Warning Signs....362
E. Identification of Supply Circuits....362

Sec. 42. General—Rules for Employees....362
420. Personal General Precautions....362
A. Rules and Emergency Methods....362
B. Qualifications of Employees....362
C. Safeguarding Oneself and Others....363
D. Energized or Unknown Conditions....363
E. Ungrounded Metal Parts....364
F. Arcing Conditions....364
G. Liquid Cell Batteries....364
H. Tools and Protective Equipment....364
I. Clothing....364
J. Ladders and Supports....364
K. Safety Straps....365
L. Fire Extinguishers....365
M. Machines and Moving Parts....365
N. Fuses....366
O. Cable Reels....366
P. Street and Area Lighting....366
421. General Operating Routines....366
A. Duties of a First Level Supervisor or Person in Charge....366
B. Area Protection....366
1. Areas Accessible to Vehicular and Pedestrian Traffic....366
2. Areas Accessible to Employees Only....367
3. Locations with Crossed or Fallen Wires....367
C. Escort....367

422. Overhead Line Operating Procedures 368
A. Setting, Moving, or Removing Poles In or Near Energized Electric Supply Lines 368
B. Checking Structures Before Climbing 368
C. Installing and Removing Wires or Cables 368
423. Underground Line Operating Procedures 369
A. Guarding Manhole and Street Openings 369
B. Testing for Gas in Manholes and Unventilated Vaults 369
C. Flames 369
D. Excavation 370
E. Identification 370
F. Operation of Power Driven Equipment 370

Sec. 43. Additional Rules for Communication Employees 371
430. General 371
431. Approach to Energized Conductors or Parts 371
432. Climbing Joint-Use Poles 371
433. Attendant on Surface at Joint-Use Manhole 372
434. Sheath Continuity 372

Sec. 44. Additional Rules for Supply Employees 372
440. General 372
441. Approach to Energized Conductors or Parts 372
A. Clearance from Live Parts 372
442. Switching Control Procedures 380
A. Designated Person 380
B. Specific Work 380
C. Operations at Stations 380
D. Re-Energizing After Work 380
E. Tagging Electric Supply Circuits 381
F. Restoration of Service After Automatic Trip 381
G. Repeating Oral Messages 381
443. Work on Energized Lines and Equipment 381
A. General Requirements 381
B. Requirement for Assisting Employee 383
C. Opening and Closing Switches 383
D. Working Position 383
E. Protecting Employees by Switches and Disconnectors 383

F. Making Connections 383
G. Switchgear 383
H. Current Transformer Secondaries 383
I. Capacitors 384
J. Gas-Insulated Equipment 384
K. Attendant on Surface 384
L. Unintentional Grounds on Delta Circuits 384
444. De-Energizing Equipment or Lines to Protect Employees 385
A. Application of Rule 385
B. Employee's Request 385
C. Operating Switches, Disconnectors, and Tagging 385
D. Employee's Protective Grounds 386
E. Proceeding with Work 386
F. Reporting Clear—Transferring Responsibility ...386
G. Removal of Tags 387
H. Sequence of Re-Energizing 387
445. Protective Grounds 387
A. Installing Grounds 387
1. Size of Grounds 387
2. Test for Voltage 388
3. Initial Connections 388
4. Completing Grounds 388
B. Removing Grounds 388
446. Live-Line Work 388
A. Training 389
B. Equipment 389
C. Bonding and Shielding for Bare-Hand Method. 389

Appendix A. Uniform System of Clearances Adopted in the 1990 Edition

Introduction 391
Clearance Rules and Tables Prior to 1990 392
Clearance Values Prior to 1990 394
Summary—Prior Editions 394
Clearances Subcommittee Activities 394
The 1990 Changes 395
Summary 397

FIGURES		RULE	PAGE
Figure D-1.	Sag and Apparent Sag	Sec. 2	58
Figure 124-1.	Clearance from Live Parts	124A1	99
Figure 124-2.	Railing Used As Guards	124C3	99
Figure 233-1.	Use of Clearance Envelope and Conductor Movement Envelopes to Determine Applicable Clearance	233A1	181
Figure 233-2.	Conductor Movement Envelope	233A1a	182
Figure 233-3.	Clearance Envelope	233A2	183
Figure 234-1 (a).	Clearance Diagram for Building	234A3	204
Figure 234-1 (b).	Clearance Diagram for Other Structures	234A3	205
Figure 234-2.	Swimming Pool Clearances	234E1	206
Figure 234-3.	Clearance Envelope for Grain Bins Filled by Portable Augurs, Conveyors, or Elevators	234E2	206
Figure 234-4.	Rail Car Clearances	234E2	207
Figure 235-1.	Clearance Diagram for Energized Conductor	235D	231
Figure 236-1.	Rule 236G, Exception 3	236G	249
Figure 237-1.	Obstruction of Working Space by Buckarm	237D1	254
Figure 250-1.	General Loading Map of United States with Respect to Loading of Overhead Lines	250B	277
Figure 250-2.	Basic Wind Speed (miles per hour)	250C	278
Figure 323-1.	Roadway Vehicle Load	323A1	339
Figure 323-2.	Wheel Load Area	323A1	339

APPENDIX FIGURES

		PAGE
Figure 1.	Basic Clearance	402
Figure 2.	Additional Clearance Required for High Temperature Operation	402
Figure 3.	Additional Clearance Required for Long Span Construction	403
Figure 4.	Clearance at Maximum Sag	403

TABLES		RULE	PAGE
Table 111-1.	Illumination Levels	111A	91
Table 124-1.	Clearance from Live Parts	124A1	100
Table 125-1.	Working Space	125A3	107
Table 127-1.	Electrical Equipment Classified Areas—Flammable Liquid Storage Areas	127C1	116
Table 127-2.	Electrical Equipment Classified Areas—Bulk Plants	127D	119
Table 127-3.	Electrical Equipment Classified Areas—Gasoline Dispensing Stations	127E3	119
Table 127-4.	Electrical Equipment Classified Areas—LPG Storage	127K3	122
Table 127-5.	Electrical Equipment Classified Areas—Natural Gas (Methane) Areas	127L3	124
Table 232-1.	Vertical Clearance of Wires, Conductors, and Cables Above Ground, Roadway, Rail, or Water Surfaces	232B1	162
Table 232-2.	Vertical Clearance of Equipment Cases and Unguarded Rigid Live Parts Above Ground or Roadway Surfaces	232B2	170
Table 232-3.	Reference Heights	232D2	174
Table 232-4.	Electrical Component of Clearance in Rule 232D3a	232D3a	175
Table 233-1.	Vertical Clearance Between Wires, Conductors, and Cables Carried on Different Supporting Structures	233C1	184
Table 233-2.	Clearance Between Supply Wires, Conductors, and Cables in Rule 233C3b(1)	233C3b(1)	191
Table 233-3.	Reference Heights	233C3a	193
Table 234-1.	Clearance of Wires, Conductors, Cables, and Unguarded Rigid Live Parts Adjacent but Not Attached to Buildings and		

TABLES		RULE	PAGE
	Other Installations Except Bridges	234E1	208
Table 234-2.	Clearance of Supply Conductors, Cables, and Unguarded Rigid Live Parts from Bridges	234E1	212
Table 234-3.	Clearance of Wires, Conductors, and Cables Passing Over or Near Swimming Areas	234E1	216
Table 234-4.	Electrical Component of Clearance to Buildings, Bridges, and Other Installations in Rule 234H3a	234H3	218
Table 234-5.	Reference Distances	234H2	219
Table 235-1.	Horizontal Clearance Between Wires, Conductors, or Cables at Supports	235B1a	232
Table 235-2.	Horizontal Clearances Between Line Conductors Smaller Than AWG No 2 at Supports, Based on Sags	235B1b(1)	233
Table 235-3.	Horizontal Clearances Between Line Conductors AWG No 2 or Larger at Supports, Based on Sags	235B1b(2)	233
Table 235-4.	Electrical Clearances in Rule 235B3a(1)	235B3a(1)	235
Table 235-5.	Vertical Clearance Between Conductors at Supports	235C1	236
Table 235-6.	Clearance in Any Direction from Line Conductors to Supports and to Vertical or Lateral Conductors, Span, or Guy Wires Attached to the Same Support	235E1	240
Table 235-7.	Clearance in Any Direction from Line Conductors to Supports	235E3b(1)(a)	245
Table 236-1.	Horizontal Clearance Between Conductors Bounding the		

TABLES		RULE	PAGE
	Climbing Space	236E	250
Table 238-1.	Vertical Clearance Between Supply Conductors and Communication Equipment, Between Communication Conductors and Supply Equipment, and Between Supply and Communication Equipment	238B	255
Table 238-2.	Vertical Clearance of Span Wires and Brackets from Communications Lines	238C	256
Table 239-1.	Clearance of Vertical and Lateral Conductors	239E1	264
Table 239-2.	Clearance Between Open Vertical Conductors and Pole Center	239E2d(1)	264
Table 241-1.	Grades of Construction for Communication Conductors Crossing Over Railroad Tracks and Supply Lines	241C3b	267
Table 242-1.	Grades of Construction for Supply Conductors Alone, at Crossing, or on the Same Structures with Other Conductors	242F	270
Table 242-2.	Grades of Construction for Communication Conductors Alone, or in Upper Position of Crossing or on Joint Poles	242F	273
Table 250-1.	Ice, Wind and Temperature	250B	279
Table 250-2.	Horizontal Wind Pressures on Cylindrical Surfaces	250C	279
Table 251-1.	Temperatures and Constants	251B2	281
Table 261-1.	Overload Capacity Factors for Reinforced Concrete Structures (Not Prestressed)	261A1a	301
Table 261-2.	Overload Capacity Factors for Metal and Prestressed		

TABLES		RULE	PAGE
	Concrete Structures	261A1	302
Table 261-3A.	Overload Capacity Factors for Wood Structures	261A2a(4)	303
Table 261-3B.	Strength Factors and Overload Capacity Factors for Wood Structures When Installed	261A2b(1)(a)	304
Table 261-4.	Overload Capacity Factors for Guys, Guy Anchors, Foundations, and Settings	261B	305
Table 261-5.	Table 261-5 not used in this edition		
Table 261-6.	Dimensions of Crossarm Cross Section of Selected Southern Pine and Douglas Fir	261D1c	306
Table 261-7.	Conductor Sizes	261H1	306
Table 262-1A.	Overload Capacity Factors for Unguyed Wood Poles	262A1b	313
Table 262-1B.	Strength Factor and Overload Capacity Factors for Unguyed Wood Poles When Installed	262A2a	314
Table 262-2.	Overload Capacity Factors for Guys	262C4	315
Table 262-3.	Overload Capacity Factors for Guys at Angles in the Line	262C4	315
Table 262-4.	Dimensions of Crossarm Cross Section of Southern Pine and Douglas Fir	262D1	315
Table 262-5.	Wire Sizes with Respect to Loading District and Span Length	262H1a	316
Table 263-1.	Sizes for Grade N Supply Line Conductors (AWG for Copper and Aluminum, Stl WG for Steel)	263D1	319
Table 263-2.	Sizes of Service Drops of 750 V or Less	263E1a(1)	319
Table 273-1.	Insulation Level Requirements	273	322
Table 341-1.	Clearance Between Supply and Communications Facilities in		

Contents

TABLES		RULE	PAGE
	Joint-Use Manholes and Vaults	341B2b(5)	343
Table 353-1.	Supply Cable or Conductor Burial Depth	353D	349
Table 431-1.	Overhead Supply Lines and Equipment Approach Distances to Exposed Energized Parts	430	371
Table 441-1.	AC Live-Line Work Minimum Clearance	441A1	373
Table 442-2.	AC Live-Line Work Minimum Clearance with Surge Factor	441A1	374
Table 441-3.	DC Minimum Clearance for Live Parts	441A1	378

APPENDIX TABLES

		PAGE
Table A-1.		398
Table A-2a.	Reference Components of Rule 232	400
Table A-2b.	Reference Components of Rule 234	401

Letter Symbols for Units

This code uses standard symbols for units. They have the following meanings:

A	ampere
ft	foot
in	inch
kV	kilovolt (1000 volts)
V	volt

Substantive changes incorporated in the 1990 edition are identified by a bar in the left-hand margin. In several cases, rules have been relocated without substantive changes in the wording. In these cases, only the rule numbers have been indicated as having been changed.

Section 1. Introduction to the National Electrical Safety Code

010. Purpose

The purpose of these rules is the practical safeguarding of persons during the installation, operation, or maintenance of electric supply and communication lines and associated equipment.

These rules contain the basic provisions that are considered necessary for the safety of employees and the public under the specified conditions. This code is not intended as a design specification or an instruction manual.

011. Scope

These rules cover supply and communication lines, equipment, and associated work practices employed by an electric supply, communication, railway, or similar utility in the exercise of its function as a utility. They cover similar systems under the control of qualified persons, such as those associated with an industrial complex or utility interactive system.

The National Electrical Safety Code (NESC) covers utility facilities and functions from the point of generation by the utility, or delivery from another entity, of electricity or communication signals through the utility system to the point of delivery to a customer's facilities.

NESC rules do not cover installations in mines, ships, railway rolling equipment, aircraft, or automotive equipment, or utilization wiring except as covered in Parts 1 and 3.

012. General Rules

All electric supply and communication lines and equipment shall be designed, operated, constructed, and maintained to meet the requirements of these rules. For all particulars not specified in these rules, construction, operation, and maintenance should be done in accordance with good practice for the given local conditions.

013. Application

A. New Installations and Extensions

1. These rules shall apply to all new installations and extensions, except that they may be waived or modified by the administrative authority. When so waived or modified, equivalent safety shall be provided in other ways, including special work methods.
2. Types of construction and methods of installation other than those specified in the rules may be used experimentally to obtain information, if done where qualified supervision is provided.

B. Existing Installations

1. Where an existing installation meets, or is altered to meet, these rules, such installation is considered to be in compliance with this edition and is not required to comply with any previous edition.
2. Existing installations, including maintenance replacements, which comply with prior editions of the code, need not be modified to comply with these rules except as may be required for safety reasons by the administrative authority.
3. Where conductors or equipment are added, altered, or replaced on an existing structure, the structure or the facilities on the structure need not be modified or replaced if the resulting installation will be in compliance with either (a) the rules that were in effect at the time of the original installation, or (b) the rules in effect in a subsequent edition to which the installation has been previously brought into compliance, or (c) the rules of this edition in accordance with Rule 013B1.

014. Waiver

The person responsible for an installation may modify or waive rules in the case of emergency or temporary installations.

A. Emergency Installations

1. The clearances required in Section 23 may be decreased for emergency installations if the following conditions are met:

a. Open supply conductors of 0–750 V, supply cables meeting Rule 230C, communication conductors and cables, guys, messengers, and neutral conductors meeting Rule 230E1 shall be suspended not less than 15.5 ft (4.8 m) above areas where vehicles are expected, or 9 ft (2.70 m) above areas where vehicles are not expected during the emergency, unless Section 23 permits lesser clearances.

b. Vertical clearances of open supply conductors above 750 V shall be increased above the applicable value of Rule 014A1 as appropriate for the voltage involved and the given local conditions.

c. Reductions in horizontal clearances permitted by this rule shall be in accordance with accepted good practice for the given local conditions during the term of the emergency.

d. Supply cables meeting Rule 230C and communication cables may be laid directly on grade if they are guarded or otherwise located so they do not obstruct pedestrian or vehicular traffic and are appropriately marked.

2. The strength of material and construction for emergency installations shall be not less than that required for Grade N construction. See Rule 263.

3. Emergency installations shall be removed, replaced, or relocated, as desired, as soon as practical.

B. Temporary Overhead Installations

When an installation is temporary, or where facilities are temporarily relocated to facilitate other work, the installation shall meet the requirements for non-temporary installation except that the strength of material and construction shall be not less than that required for Grade N construction. See Rule 263.

015. Intent

A. The word “shall” indicates provisions that are mandatory.

B. The word “should” indicates provisions that are normally and generally practical for the specified conditions. However, where the word “should” is used, it is recognized that, in certain instances, additional local conditions not specified herein may make these provisions impractical. When this occurs, the difference in conditions shall be appropriately recognized and Rule 012 shall be met.

C. The word *"RECOMMENDATION"* indicates provisions considered desirable, but that are not intended to be mandatory.

D. The word *"NOTE"* or the word *"EXAMPLE"* used in a rule indicates material provided for information or illustrative purposes only. *"NOTES"* and *"EXAMPLES"* are not mandatory and are not considered to be a part of code requirements.

E. Footnotes to a table have the force and effect required or allowed by the rule that specifies the use of the table.

F. The text of a rule is indented from its identifying numbers and letters. A *"RECOMMENDATION"* or *"EXCEPTION"* is considered to be a part of the text and to apply to all text in that rule above its location that is indented to the same level.

A *"RECOMMENDATION," "EXCEPTION,"* or *"NOTE"* applies to all text in that rule above its location that is indented to the same level.

016. Effective Date

Unless otherwise stipulated by the administrative authority, this code shall become effective 180 days following the date of its publication and shall apply to new installations and extensions where both design and approval were started after the expiration of that period. This edition may be used on or after the publication date.

NOTE: A period of 180 days is allowed for utilities and regulatory authorities to acquire copies of the new edition and to change regulations, internal standards and procedures as may be required.

017. Units of Measure

Numerical values in the requirements of this code are stated in the customary inch-foot-pound system and in the metric system (SI).[1] In text the customary inch-foot-pound system value is shown first with the metric value (inside parentheses)

[1] Le Système Internationale d'Unités (The International System of Units [or SI]) is the modern version of the metric system. For basic information and conversion factors see ANSI/IEEE Std 268-1982 [28], listed in Section 3.

following. Extensive detailed tables are duplicated. The first, marked IN, FT, or LB, contains the inch-foot-pound values; the second, marked M, contains metric (SI) values. Tensions are stated in newtons, the SI unit of force.

The SI values and the customary inch-foot-pound values are not, nor are they intended to be, identical measures. The values shown in each system of measurement have been rounded to convenient numbers in order to simplify measurement and minimize errors. The values shown in each system are functional equivalents for safety purposes.

The values required in this code have been carefully developed and evaluated to assure that the intended levels of safety are provided in both systems; neither is distinguishable from the other for safety purposes. The values specified in either system of measurement may be used, or the values of the two systems may be intermixed, as desired.[2]

---

[2]It is recognized that many equivalent utility system components may be purchased in both SI and customary units.

Section 2. Definitions of Special Terms

The following definitions are for use with the National Electrical Safety Code.

For other use, and for definitions not contained herein, see ANSI/IEEE Std 100-1988 [27].[3]

administrative authority. The governmental authority exercising jurisdiction over application of this code.

alive or live. *See* **energized.**

automatic. Self-acting, operating by its own mechanism when actuated by some impersonal influence—as, for example, a change in current strength; not manual; without personal intervention. Remote control that requires personal intervention is not automatic, but manual.

backfill (noun). Materials such as sand, crushed stone, or soil, which are placed to fill an excavation.

ballast section (railroads). The section of material, generally trap rock, which provides support under railroad tracks.

bonding. The electrical interconnecting of conductive parts, designed to maintain a common electrical potential.

cable. A conductor with insulation, or a stranded conductor with or without insulation and other coverings (single-conductor cable) or a combination of conductors insulated from one another (multiple-conductor cable).

spacer cable. A type of electric supply line construction consisting of an assembly of one or more covered conductors, separated from each other and supported from a messenger by insulating spacers.

cable jacket. A protective covering over the insulation, core, or sheath of a cable.

---

[3]The numbers in brackets correspond to those of the references listed in Section 3.

cable sheath. A conductive protective covering applied to cables. *NOTE:* A cable sheath may consist of multiple layers, of which one or more is conductive.

cable terminal. A device which provides insulated egress for the conductors. *Syn:* **termination.**

circuit. A conductor or system of conductors through which an electric current is intended to flow.

circuit breaker. A switching device capable of making, carrying and breaking currents under normal circuit conditions and also making, carrying for a specified time, and breaking currents under specified abnormal conditions such as those of short circuit.

clearance. The clear distance between two objects measured surface-to-surface.

common use. Simultaneous use by two or more utilities of the same kind.

communication lines. See **lines, communication.**

conductor.

1. A material, usually in the form of a wire, cable, or bus bar, suitable for carrying an electric current.
2. **bundled conductor.** An assembly of two or more conductors used as a single conductor and employing spacers to maintain a predetermined configuration. The individual conductors of this assembly are called subconductors.
3. **covered conductor.** A conductor covered with a dielectric having no rated insulating strength or having a rated insulating strength less than the voltage of the circuit in which the conductor is used.
4. **fiber optic conductor.** See **fiber optic cable—communication** or **fiber optic cable—supply.**
5. **grounded conductor.** A conductor which is intentionally grounded, either solidly or through a noninterrupting current-limiting device.
6. **grounding conductor.** A conductor which is used to connect the equipment or the wiring system with a grounding electrode or electrodes.

7. **insulated conductor.** A conductor covered with a dielectric (other than air) having a rated insulating strength equal to or greater than the voltage of the circuit in which it is used.

8. **lateral conductor.** A wire or cable extending in a general horizontal direction at an angle to the general direction of the line conductors, and entirely supported on one structure.

9. **line conductor.** (Overhead supply or communication lines). A wire or cable intended to carry electric currents, extending along the route of the line, supported by poles, towers or other structures, but not including vertical or lateral conductors.

10. **open conductor.** A type of electric supply or communication line construction in which the conductors are bare, covered or insulated and without grounded shielding, individually supported at the structure either directly or with insulators. *Syn:* **open wire.**

conductor shielding. An envelope which encloses the conductor of a cable and provides an equipotential surface in contact with the cable insulation.

conduit. A structure containing one or more ducts.
NOTE: Conduit may be designated as iron pipe conduit, tile conduit, etc. If it contains one duct only it is called *single-duct conduit;* if it contains more than one duct it is called *multiple-duct conduit,* usually with the number of ducts as a prefix, for example, *two-duct multiple conduit.*

conduit system. Any combination of duct, conduit, conduits, manholes, handholes and vaults joined to form an integrated whole.

current-carrying part. A conducting part intended to be connected in an electric circuit to a source of voltage. Non-current-carrying parts are those not intended to be so connected.

de-energized. Free from any electrical connection to a source of potential difference and from electric charge; not having a potential different from that of the earth.
NOTE: The term is used only with reference to current-carrying parts which are sometimes energized (alive). *Syn:* **dead.**

designated person. A qualified person designated to perform specific duties under the conditions existing. *Syn:* **designated employee.**

disconnecting or isolating switch. A mechanical switching device used for changing the connections in a circuit, or for isolating a circuit or equipment from a source of power.
NOTE: It is required to carry normal load current continuously, and also abnormal or short-circuit current for short intervals as specified. It is also required to open or close circuits either when negligible current is broken or made, or when no significant change in the voltage across the terminals of each of the switch poles occurs. *Syn:* **disconnector, isolator.**

duct. A single enclosed raceway for conductors or cable.

effectively grounded. Intentionally connected to earth through a ground connection or connections of sufficiently low impedance and having sufficient current-carrying capacity to prevent the buildup of voltages which may result in undue hazard to connected equipment or to persons.

electric supply equipment. Equipment which produces, modifies, regulates, controls, or safeguards a supply of electric energy. *Syn:* **supply equipment.**

electric supply lines. *See* **lines.**

electric supply station. Any building, room, or separate space within which electric supply equipment is located and the interior of which is accessible, as a rule, only to qualified persons. This includes generating stations and substations, including their associated generator, storage battery, transformer and switchgear rooms or enclosures but does not include facilities such as pad-mounted equipment and installations in manholes and vaults.

enclosed. Surrounded by case, cage, or fence designed to protect the contained equipment and minimize the possibility, under normal conditions, of dangerous approach or accidental contact by persons or objects.

energized. Electrically connected to a source of potential difference, or electrically charged so as to have a potential significantly different from that of earth in the vicinity.
NOTE: **alive** or **live.**

equipment. A general term including fittings, devices, appliances, fixtures, apparatus, and similar terms used as part of or in connection with an electric supply or communication system.

explosion-proof apparatus. Apparatus enclosed in a case that is capable of withstanding an explosion of a specified gas or vapor which may occur within it and of preventing the ignition of a specified gas or vapor surrounding the enclosure by sparks, flashes, or explosion of the gas or vapor within, and which operates at such an external temperature that a surrounding flammable atmosphere will not be ignited thereby.

exposed. Not isolated or guarded.

fiber optic cable-communication. A fiber optic cable meeting the requirements for a communication line and located in the communication space of overhead or underground facilities.

fiber optic cable-supply. A fiber optic cable located in the supply space of overhead or underground facilities.

fireproofing (of cables). The application of a fire-resistant covering.

grounded. Connected to or in contact with earth or connected to some extended conductive body that serves instead of the earth.

grounded effectively. *See* **effectively grounded.**

grounded system. A system of conductors in which at least one conductor or point is intentionally grounded, either solidly or through a non-interrupting current-limiting device.

guarded. Covered, fenced, enclosed, or otherwise protected, by means of suitable covers or casings, barrier rails or screens, mats or platforms, designed to minimize the possibility, under normal conditions, of dangerous approach or accidental contact by persons or objects.

NOTE: Wires which are insulated, but not otherwise protected, are not considered as guarded.

handhole. An access opening, provided in equipment or in a below-the-surface enclosure in connection with underground lines, into which personnel reach but do not enter, for the purpose of installing, operating, or maintaining equipment or cable or both.

insulated. Separated from other conducting surfaces by a dielectric (including air space) offering a high resistance to the passage of current.
NOTE: When any object is said to be insulated, it is understood to be insulated for the conditions to which it is normally subjected. Otherwise, it is, within the purpose of these rules, uninsulated.

insulation (as applied to cable). That which is relied upon to insulate the conductor from other conductors or conducting parts or from ground.

insulation shielding. An envelope which encloses the insulation of a cable and provides an equipotential surface in contact with the cable insulation.

insulator. Insulating material in a form designed to support a conductor physically and electrically separate it from another conductor or object.

isolated. Not readily accessible to persons unless special means for access are used.

isolated by elevation. Elevated sufficiently so that persons may safely walk underneath.

isolator. See **disconnecting or isolating switch.**

jacket. A protective covering over the insulation, core, or sheath of a cable.

joint use. Simultaneous use by two or more kinds of utilities.

lines.
1. **communication lines.** The conductors and their supporting or containing structures which are used for public or private signal or communication service, and which operate at potentials not ex-

ceeding 400 volts to ground or 750 volts between any two points of the circuit, and the transmitted power of which does not exceed 150 watts. When operating at less than a nominal voltage of 90 volts, no limit is placed on the transmitted power of the system. Under specified conditions, communication cables may include communication circuits exceeding the preceding limitation where such circuits are also used to supply power solely to communication equipment.

NOTE: Telephone, telegraph, railroad-signal, data, clock, fire, police-alarms, cable television and other systems conforming with the above are included. Lines used for signaling purposes, but not included under the above definition, are considered as supply lines of the same voltage and are to be so installed.

2. **electric supply lines.** Those conductors used to transmit electric energy and their necessary supporting or containing structures. Signal lines of more than 400 volts are always supply lines within the meaning of the rules, and those of less than 400 volts may be considered as supply lines, if so run and operated throughout. *Syn:* **supply lines.**

low voltage protection. The effect of a device operative on the reduction or failure of voltage so as to cause and maintain the interruption of power supply to the equipment protected.

manhole. A subsurface enclosure which personnel may enter and which is used for the purpose of installing, operating, and maintaining submersible equipment and cable.

manhole cover. A removable lid which closes the opening to a manhole or similar subsurface enclosure.

manhole grating. A grid which provides ventilation and a protective cover for a manhole opening.

manual. Capable of being operated by personal intervention.

pad-mounted equipment. A general term describing enclosed equipment, the exterior of which enclosure is at ground potential, positioned on a surface-mounted pad.

prestressed concrete structures. Concrete structures which include metal tendons that are tensioned and anchored either before or after curing of the concrete.

pulling iron. An anchor secured in the wall, ceiling, or floor of a manhole or vault to attach rigging used to pull cable.

pulling tension. The longitudinal force exerted on a cable during installation.

qualified. Having adequate knowledge of the installation, construction or operation of apparatus and the hazards involved.

raceway. Any channel designed expressly and used solely for holding conductors.

random separation. Installed with no deliberate separation.

readily climbable. Having sufficient handholds and footholds to permit an average person to climb easily without using a ladder or other special equipment.

remotely operable (as applied to equipment). Capable of being operated from a position external to the structure in which it is installed or from a protected position within the structure.

roadway. The portion of highway, including shoulders, for vehicular use.

NOTE: A divided highway has two or more roadways. *See also* **shoulder; traveled way.**

rural districts. All places not urban. This may include thinly settled areas within city limits.

sag.

1. The distance measured vertically from a conductor to the straight line joining its two points of support. Unless otherwise stated in the rule, the sag referred to is the sag at the midpoint of the span. See Fig D-1.

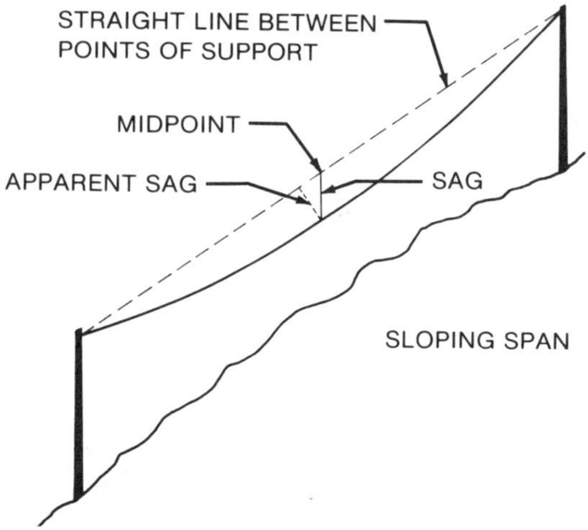

Fig D-1
Sag and Apparent Sag

2. **initial unloaded sag.** The sag of a conductor prior to the application of any external load.

3. **final sag.** The sag of a conductor under specified conditions of loading and temperature applied, after it has been subjected for an appreciable period to the loading prescribed for the loading district in which it is situated, or equivalent loading, and the loading removed. Final sag shall include the effect of inelastic deformation (creep).

4. **final unloaded sag.** The sag of a conductor after it has been subjected for an appreciable period to the loading prescribed for the loading district in which it is situated, or equivalent loading, and the loading removed. Final unloaded sag shall include the effect of inelastic deformation (creep).

5. **total sag.** The distance measured vertically from the conductor to the straight line joining its two points of support, under conditions of ice loading equivalent to the total resultant loading for the district in which it is located.

6. **maximum total sag.** The total sag at the midpoint of the straight line joining the two points of support of the conductor.

7. **apparent sag of a span.** The maximum distance between the wire in a given span and the straight line between the two points of

support of the wire, measured perpendicularly from the straight line. See Fig D-1.

8. **sag of a conductor at any point in a span.** The distance measured vertically from the particular point in the conductor to a straight line between its two points of support.

9. **apparent sag at any point in the span.** The distance, at the particular point in the span, between the wire and the straight line between the two points of support of the wire, measured perpendicularly from the straight line.

separation. The distance between two objects, measured surface-to-surface, and usually filled with a solid or liquid material.

service drop. The overhead conductors between the electric supply or communication line and the building or structure being served.

shoulder. The portion of the roadway contiguous with the traveled way for accommodation of stopped vehicles for emergency use and for lateral support of base and surface course.

side-wall pressure. The crushing force exerted on a cable during installation.

span length. The horizontal distance between two adjacent supporting points of a conductor.

span wire. An auxiliary suspension wire which serves to support one or more trolley contact conductors or a light fixture and the conductors which connect it to a supply system.

structure conflict. A line so situated with respect to a second line that the overturning of the first line will result in contact between its supporting structures or conductors and the conductors of the second line, assuming that no conductors are broken in either line.

supply equipment. *See* **electric supply equipment.**

supply station. *See* **electric supply station.**

supporting structure. The main supporting unit (usually a pole or tower).

susceptiveness. The characteristics of a communications circuit including its connected apparatus which determine the extent to which it is adversely affected by inductive fields.

switch. A device for opening and closing or for changing the connection of a circuit. In these rules, a switch is understood to be manually operable, unless otherwise stated.

switchboard. A type of switchgear assembly that consists of one or more panels with electric devices mounted thereon, and associated framework.

tag. Accident prevention tag (DANGER, PEOPLE AT WORK, etc) of a distinctive appearance used for the purpose of personnel protection to indicate that the operation of the device to which it is attached is restricted.

tension, unloaded.
1. **initial.** The longitudinal tension in a conductor prior to the application of any external load.
2. **final.** The longitudinal tension in a conductor after it has been subjected for an appreciable period to the loading prescribed for the loading district in which it is situated, or equivalent loading, and the loading removed. Final unloaded tension shall include the effect of inelastic deformation (creep).

termination. *See* **cable terminal.**

transformer vault. An isolated enclosure either above or below ground with fire-resistant walls, ceiling, and floor, in which transformers and related equipment are installed, and which is not continuously attended during operation. *See also* **vault.**

traveled way. The portion of the roadway for the movement of vehicles, exclusive of shoulders and full-time parking lanes.

urban districts. Thickly settled areas (whether in cities or suburbs) or where congested traffic often occurs. A highway, even though in thinly settled areas, on which the traffic is often very heavy, is considered as urban.

utility. An organization responsible for the installation, operation or maintenance of electric supply or communication systems.

utility interactive system. An electric power production system which is operating in parallel with and capable of delivering energy to a utility electric supply system.

utilization equipment. Equipment, devices, and connected wiring which utilize electric energy for mechanical, chemical, heating, lighting, testing, or similar purposes and are not a part of supply equipment, supply lines, or communication lines.

vault. An enclosure above or below ground which personnel may enter and is used for the purpose of installing, operating, or maintaining equipment or cable which need not be of a submersible design.

voltage.

1. The effective (rms) potential difference between any two conductors or between a conductor and ground. Voltages are expressed in nominal values unless otherwise indicated. The nominal voltage of a system or circuit is the value assigned to a system or circuit of a given voltage class for the purpose of convenient designation. The operating voltage of the system may vary above or below this value.

2. **voltage of circuit not effectively grounded.** The highest nominal voltage available between any two conductors of the circuit.

NOTE: If one circuit is directly connected to and supplied from another circuit of higher voltage (as in the case of an autotransformer), both are considered as of the higher voltage, unless the circuit of the lower voltage is effectively grounded, in which case its voltage is not determined by the circuit of higher voltage. Direct connection implies electric connection as distinguished from connection merely through electromagnetic or electrostatic induction.

3. **voltage of a constant current circuit.** The highest normal full load voltage of the current.

4. **voltage of an effectively grounded circuit.** The highest nominal voltage available between any conductor of the circuit and ground unless otherwise indicated.

5. **voltage to ground of:**

a. **a grounded circuit.** The highest nominal voltage available between any conductor of the circuit and that point or conductor of the circuit which is grounded.

b. **an ungrounded circuit.** The highest nominal voltage available between any two conductors of the circuit concerned.

6. **voltage to ground of a conductor of:**

a. **a grounded circuit.** The nominal voltage between such conductor and that point or conductor of the circuit which is grounded.

b. **an ungrounded circuit.** The highest nominal voltage between such conductor and any other conductor of the circuit concerned.

wire gages. Throughout these rules the American Wire Gage (AWG), formerly known as Brown & Sharpe (B&S), is the standard gage for copper, aluminum and other conductors, excepting only steel conductors for which the Steel Wire Gage (Stl WG) is used.
NOTE: The Birmingham Wire Gage is obsolete.

Section 3. References

The following standards form a part of the National Electrical Safety Code to the extent indicated in the rules herein.

[1] ANSI A12.1-1973, Safety Requirements for Floor and Wall Openings, Railings, and Toeboards.[4]

[2] ANSI A14.1-1982, Safety Requirements for Portable Wood Ladders.

[3] ANSI A14.2-1982, Safety Requirements for Portable Metal Ladders.

[4] ANSI A14.3-1984, Safety Requirements for Fixed Ladders.

[5] ANSI A14.5-1982, Safety Requirements for Portable Reinforced Plastic Ladders.

[6] ANSI A58.1-1982, Building Code Requirements for Minimum Design Loads in Buildings and Other Structures.

[7] ANSI A92.2-1979, Vehicle Mounted Elevating and Rotating Aerial Devices.

[8] ANSI C29.1-1982, Test Methods for Electrical Power Insulators.

[9] ANSI C29.2-1983, Wet-Process Porcelain and Toughened Glass Insulators (Suspension Type).

[10] ANSI C29.3-1986, Wet-Process Porcelain Insulators (Spool Type).

[11] ANSI C29.4-1984, Wet-Process Porcelain Insulators (Strain Type).

---

[4]ANSI publications are available from the Sales Department, American National Standards Institute, 1430 Broadway, New York, NY 10018.

[12] ANSI C29.5-1984, Low- and Medium-Voltage Pin Type Wet-Process Porcelain Insulators.

[13] ANSI C29.6-1984, High-Voltage Pin Type Wet-Process Porcelain Insulators.

[14] ANSI C29.7-1983, High-Voltage Line-Post Type Wet-Process Porcelain Insulators.

[15] ANSI C84.1-1982, Voltage Ratings for Electric Power Systems and Equipment (60 Hz).

[16] ANSI C92.1-1982, Insulation Coordination.

[17] ANSI O5.1-1987, Specifications and Dimensions for Wood Poles.

[18] ANSI Z53.1-1979, Safety Color Code for Marking Physical Hazards.

[19] ANSI Z87.1-1979, Practice for Occupational and Educational Eye and Face Protection.

[20] ANSI Z88.2-1980, Practices for Respiratory Protection.

[21] ANSI Z89.1-1986, Protective Headwear for Industrial Workers.

[22] ANSI Z244.1-1982, Safety Requirements for the Lockout/Tagout of Energy Sources.

[23] ANSI/ASME B15.1-1984, Safety Standard for Mechanical Power Transmission Apparatus.

[24] ANSI/ASTM D1050-80, Specifications for Rubber Insulating Line Hose.

[25] ANSI/ASTM F496-80, Specifications for In-Service Care of Insulating Gloves and Sleeves.

[26] ANSI/IEEE Std 80-1986, IEEE Guide for Safety in AC Substation Grounding.[5]

[27] ANSI/IEEE Std 100-1988, IEEE Standard Dictionary of Electrical and Electronics Terms.

[28] ANSI/IEEE Std 268-1982, IEEE Standard Metric Practice.

[29] ANSI/IEEE Std 1119-1988, IEEE Guide for Fence Safety Clearances in Electric Supply Stations.

[30] ANSI/NFPA 10-1988, Portable Fire Extinguisher.

[31] ANSI/NFPA 30-1987, Flammable and Combustible Liquids Code.

[32] ANSI/NFPA 70-1987, National Electrical Code.[6]

[33] ANSI/NFPA 77-1983, Recommended Practice on Static Electricity.

[34] ANSI/NFPA 85F-1982, Standard for the Installation and Operation of Pulverized Fuel Systems.

[35] API RP500A (822-50000), Recommended Practice for Classification of Areas for Electrical Installations in Petroleum Refineries.[7]

[36] ASTM D120-84a, Specification for Rubber Insulating Gloves.[8]

---

[5]ANSI/IEEE standards are available from Publication Sales, IEEE Service Center, 445 Hoes Lane, Piscataway, NJ 08854, and from the Sales Department, American National Standards Institute, 1430 Broadway, New York, NY 10018.

[6]The National Electrical Code is published by the National Fire Protection Association, Batterymarch Park, Quincy, MA 02269. Copies are also available from the Sales Department, American National Standards Institute, 1430 Broadway, New York, NY 10018.

[7]API publications are available from the American Petroleum Institute (API), 2101 L Street, NW, Washington, DC 20037.

[8]ASTM publications are available from American Society for Testing and Materials, 1916 Race Street, Philadelphia, PA 19103.

[37] ASTM D178-81, Specification for Rubber Insulating Matting.

[38] ASTM D1048-81, Specification for Rubber Insulating Blankets.

[39] ASTM D1049-83, Specification for Rubber Insulating Covers.

[40] ASTM D1051-81, Specification for Rubber Insulating Sleeves.

[41] ASTM F478-81, Specification for In-Service Care of Insulating Line Hose and Covers.

[42] ASTM F479-83, Specification for In-Service Care of Insulating Blankets.

[43] NFPA 496-1982, Standard for Purged and Pressurized Enclosures for Electrical Equipment in Hazardous (Classified) Locations.[9]

[9]NFPA publications are available from the National Fire Protection Association, Batterymarch Park, Quincy, MA 02269.

Section 9. Grounding Methods for Electric Supply and Communication Facilities

90. Purpose

The purpose of Section 9 of this code is to provide practical methods of grounding, as one of the means of safeguarding employees and the public from injury that may be caused by electrical potential.

91. Scope

Section 9 of this code covers methods of protective grounding of supply and communication conductors and equipment. The rules requiring grounding are in other parts of this code.

These rules do not cover the grounded return of electric railways nor those lightning protection wires which are normally independent of supply or communication wires or equipment.

92. Point of Connection of Grounding Conductor

A. Direct Current Systems Which Are To Be Grounded

1. 750 volts and below

Connection shall be made only at supply stations. In three-wire direct-current systems the connection shall be made to the neutral.

2. Over 750 volts

Connection shall be made at both the supply and load stations. The connection shall be made to the neutral of the system. The ground or grounding electrode may be external to or remotely located from each of the stations.

One of the two stations may have its ground connection made through surge arresters provided the other station neutral is effectively grounded as described above.

B. Alternating Current Systems Which Are To Be Grounded

1. 750 volts and below

The point of the grounding connection on a wye-connected three-phase four-wire system, or on a single-phase three-wire system, shall be the neutral conductor. On other one-, two-, or three-phase systems with an associated lighting circuit or circuits, the point of grounding connection shall be on the common circuit conductor associated with the lighting circuits.

The point of grounding connection on a three-phase three-wire system, whether derived from a delta-connected or an ungrounded wye-connected transformer installation not used for lighting, may be any of the circuit conductors, or it may be a separately derived neutral.

The grounding connections shall be made at the source, and at the line side of all service equipment.

2. Over 750 volts
   a. Nonshielded (Bare or Covered Conductors or Insulated Nonshielded Cables)
      Grounding connection shall be made at the neutral of the source. Additional connections may be made, if desired, along the length of the neutral, where this is one of the system conductors.
   b. Shielded
      (1) Surge Arrester Cable-Shielding Interconnection
          Cable shielding grounds shall be bonded to surge arrester grounds, where provided, at points where underground cables are connected to overhead lines.
      (2) Cable Without Insulating Jacket
          Connection shall be made to the neutral of the source transformer and at cable termination points.
      (3) Cable With Insulating Jacket
          Additional bonding and connections between the cable insulation shielding or sheaths and the system ground are recommended. In multi-grounded shielded cable systems, the shielding (including sheath) shall be grounded at each cable joint exposed to personnel contact. Where multigrounded shielding cannot be used for electrolysis or sheath current reasons, the shielding sheaths and splice enclosure devices shall be insulated for the voltage that may appear on them during normal operation.

Bonding transformers or reactors may be substituted for direct ground connection at one end of the cable.

3. Separate Grounding Conductor

If a separate grounding conductor is used as an adjunct to a cable run underground, it shall be connected either directly or through the system neutral to the source transformer and accessories and cable accessories where these are to be grounded. This grounding conductor shall be located in the same direct burial or duct bank run (or the same duct if this is of magnetic material) as the circuit conductors.

EXCEPTION: The grounding conductor for a circuit which is installed in a magnetic duct need not be in the same duct if the duct containing the circuit is bonded to the separate grounding conductor at both ends.

C. Messenger Wires and Guys

1. Messenger Wires

Messenger wires required to be grounded shall be connected to grounding conductors at poles or structures at maximum intervals as listed below.

a. Where messenger wires are adequate for system grounding conductors (Rules 93C1, 93C2, and 93C5), four connections in each mile (1.6 km).

b. Where messenger wires are not adequate for system grounding conductors, eight connections per mile (1.6 km), exclusive of service grounds.

2. Guys

Guys which are required to be grounded shall be connected to one or more of the following:

a. A grounded metallic supporting structure.

b. An effective ground on a nonmetallic supporting structure.

c. A line conductor which has at least four ground connections in each mile of line in addition to the ground connections at individual services.

3. Common Grounding of Messengers and Guys on the Same Supporting Structure

a. Where messengers and guys on the same supporting structure are required to be grounded, they shall be bonded together and grounded by connection to:

(1) One grounding conductor that is grounded at that structure, or to

(2) Separate grounding conductors or grounded messengers that are bonded together and grounded at that structure, or to

(3) One or more grounded line conductors or grounded messengers that are (a) bonded together at this structure or elsewhere and (b) multigrounded elsewhere at intervals as specified in Rules 92C1 and 92C2.

b. At common crossing structures, messengers and guys that are required to be grounded shall be bonded together at that structure and grounded in accordance with Rule 92C3a.

EXCEPTION: This rule does not apply to guys that are connected to an effectively grounded overhead static wire.

D. Current in Grounding Conductor

Ground connection points shall be so arranged that under normal circumstances there will be no objectionable flow of current over the grounding conductor. If an objectionable flow of current occurs over a grounding conductor due to the use of multiple grounds, one or more of the following should be used:

1. Abandon one or more grounds.
2. Change location of grounds.
3. Interrupt the continuity of the conductor between ground connections.
4. Subject to the approval of the administrative authority, take other effective means to limit the current.

The system ground of the source transformer shall not be removed.

The temporary currents set up under abnormal conditions while the grounding conductors are performing their intended protective functions are not considered objectionable. The conductor shall have the capability of conducting anticipated fault current without thermal overloading or excessive voltage buildup. Refer to Rule 93C.

E. Fences

Fences, where required to be grounded by other parts of this code, shall be grounded at or near the location of a supply line or lines crossing them, and additionally, at distances not exceeding 150 ft (45 m) on either side. Fences shall also be grounded at each side of a gate or other opening in the fence. Any gate or other opening shall also be bonded across by a buried bonding jumper. A gate shall be metallically connected or bonded to the grounding conductor, jumper, or fence. Separate barbed wire strands above fencing, on non-conducting posts, shall be bonded to metallic fencing or grounding conductors at the grounding points.

Where required to be grounded, fences shall be bonded to the grounding system of the enclosed equipment or to a separate underground conductor below or near the fence line.

93. Grounding Conductor and Means of Connection

A. Composition of Grounding Conductors

In all cases the grounding conductor shall be made of copper or other metals or combinations of metals which will not corrode excessively during the expected service life under the existing conditions and, if practical, shall be without joint or splice. If joints are unavoidable, they shall be so made and maintained as to not materially increase the resistance of the grounding conductor and shall have appropriate mechanical and corrosion-resistant characteristics. For surge arresters and ground detectors, the grounding conductor or conductors shall be as short, straight, and free from sharp bends as practical. The structural metal frame of a building or structure may serve as a grounding conductor to an acceptable grounding electrode.

In no case shall a circuit-opening device be inserted in the grounding conductor or connection except where its operation will result in the automatic disconnection from all sources of energy of the circuit leads connected to the equipment so grounded.

EXCEPTION 1: Temporary disconnection of grounding conductors for testing purposes, under competent supervision, shall be permitted.

EXCEPTION 2: Disconnection of a grounding conductor from a surge arrester is allowed when accomplished by means of a surge arrester disconnector.

NOTE: The base of the surge arrester may remain at line potential following operation of the disconnector.

B. Connection of Grounding Conductors

Connection of the grounding conductor shall be made by a means matching the characteristics of both the grounded and grounding conductors, and suitable for the environmental exposure. These means include brazing, welding, mechanical and compression connections, ground clamps, and ground straps. Soldering is acceptable only in conjunction with lead sheaths.

C. Ampacity and Strength

The short time ampacity of a bare grounding conductor is that current which the conductor can carry for the time during which the current flows without melting or separating under the applied tensions. If a grounding conductor is insulated, its short time ampacity is the current which it can carry for the applicable time without damaging the insulation. Where grounding conductors at one location are paralleled, the increased total current capacity may be considered.

1. System Grounding Conductors for Single-Grounded Systems

   The system grounding conductor or conductors for a system with single system grounding electrode or set of electrodes, exclusive of grounds at individual services, shall have a short time ampacity adequate for the fault current which can flow in the grounding conductor or conductors for the operating time of the system protective device. If this value cannot be readily determined, continuous ampacity of the grounding conductor or conductors shall be not less than the full load continuous current of the system supply transformer or other source of supply.

2. System Grounding Conductors for Multigrounded Alternating Current Systems

   The system grounding conductors for an alternating current system with grounds at more than one location exclusive of grounds at individual services shall have continuous total ampacities at each location of not less than one-fifth that of the conductors to which they are attached. (See also Rule 93C8).

3. Grounding Conductors for Instrument Transformers
The grounding conductor for instrument cases and secondary circuits for instrument transformers shall not be smaller than AWG No 12 copper or shall have equivalent short time ampacity.
4. Grounding Conductors for Primary Surge Arresters
The grounding conductor or conductors shall have adequate short time ampacity under conditions of excess current caused by or following a surge. Individual arrester grounding conductors shall be no smaller than AWG No 6 copper or AWG No 4 aluminum.
EXCEPTION: Arrester grounding conductors may be copper-clad or aluminum-clad steel wire having not less than 30% of the conductivity of solid copper or aluminum wire of the same diameter.
Where flexibility of the grounding conductor, such as adjacent to the base of the arrester, is vital to its proper operation, a suitably flexible conductor shall be employed.
5. Grounding Conductors for Equipment, Messenger Wires, and Guys
   a. Conductors
   The grounding conductors for equipment, raceways, cable, messenger wires, guys, sheaths, and other metal enclosures for wires shall have short time ampacities adequate for the available fault current and operating time of the system fault protective device. If no overcurrent or fault protection is provided, the ampacity of the grounding conductor shall be determined by the design and operating conditions of the circuit, but shall not be less than that of AWG No 8 copper. Where the adequacy and continuity of the conductor enclosures and their attachment to the equipment enclosures is assured, this path can constitute the equipment grounding conductor.
   b. Connections
   Connections of the grounding conductor shall be to a suitable lug, terminal, or device not disturbed in normal inspection, maintenance, or operation.

6. Fences
The grounding conductor for fences required to be grounded by other parts of this code shall be any of those meeting the requirements of Rule 93C5 or shall be steel wire not smaller than Stl WG No 5. It shall be connected to the fence posts with connecting means suitable for the material when the posts are of conducting material. If the posts are of nonconducting material, suitable bonding connections shall be made to the fence mesh strands and the barbed wire strands at each grounding conductor point.
7. Bonding of Equipment Frames and Enclosures
Where required, a low impedance metallic path shall be provided for the passage of possible conductor or equipment, or both, fault current back to the grounded terminal of the supply, where the supply is local. Where the supply is remote, the metallic path shall interconnect the equipment frames and enclosures with all other nonenergized conducting components within reach and shall additionally be connected to ground as outlined in Rule 93C5. Short-time ampacities of bonding conductors shall be adequate for the duty involved.
8. Ampacity Limit
No grounding conductor need have greater ampacity than either:
    a. The phase conductors which would supply the ground fault current, or
    b. The maximum current which can flow through it to the ground electrode or electrodes to which it is attached. For a single grounding conductor and connected electrode or electrodes, this would be the supply voltage divided by the electrode resistance (approximately).
9. Strength
All grounding conductors shall have mechanical strength suitable for the conditions to which they may reasonably be subjected.

Further, unguarded grounding conductors shall have a tensile strength not less than that of AWG No 8 soft-drawn copper, except as noted in Rule 93C3.

D. Guarding and Protection

1. The grounding conductors for single-grounded systems and those exposed to mechanical damage shall be guarded. However, grounding conductors need not be guarded where not readily accessible to the public nor where grounding multigrounded circuits or equipment.
2. Where guarding is required, grounding conductors shall be protected by guards suitable for the exposure to which they may reasonably be subjected. The guards should extend for not less than 8 ft (2.45 m) above the ground or platform from which the grounding conductors are accessible to the public.
3. Where guarding is not required, grounds shall be protected by being substantially attached closely to the surface of the pole or other structure in areas of exposure to mechanical damage and, where practical, on the portion of the structure having least exposure.
4. Guards used for grounding conductors of lightning protection equipment shall be of nonmetallic materials if the guard completely encloses the grounding conductor or is not bonded at both ends to the grounding conductor.

E. Underground

1. Grounding conductors laid directly underground shall be laid slack or shall be of sufficient strength to prevent being readily broken by earth movement or settling normal at the particular location.
2. Direct-buried uninsulated joints or splices in grounding conductors should be welded, brazed, or of the compression type to minimize the possibility of loosening or corrosion. The number of joints or splices should be the minimum practical.
3. Grounding cable insulation shielding systems shall be interconnected with all other accessible grounded power supply equipment in manholes, handholes, and vaults.
   EXCEPTION: Where cathodic protection or shield cross-bonding is involved, interconnection may be omitted.
4. Looped magnetic elements such as structural steel, piping, reinforcing bars, etc, should not separate grounding conductors from the phase conductors of circuits they serve.

5. Metals used for grounding, in direct contact with earth, concrete, or masonry, shall have been proven suitable for such exposure.

   NOTE 1: Under present technology, aluminum has not generally been proven suitable for such use.

   NOTE 2: Metals of different galvanic potentials which are electrically interconnected may require protection against galvanic corrosion.

6. Sheath Transposition Connections (Cross-Bonding)
   a. Where cable insulating shields or sheaths, which are normally connected to ground, are insulated from ground to minimize shield circulating currents, they shall be insulated from personnel contact at accessible locations. Transposition connections and bonding jumpers shall be insulated for nominal 600 volts service, unless the normal shielding voltage exceeds this level, in which case the insulation shall be ample for the working voltage to ground.
   b. Bonding jumpers and connecting means shall be sized and selected to carry the available fault current without damaging jumper insulation or sheath connections.

F. Common Grounding Conductor for Circuits, Metal Raceways, and Equipment

Where the ampacity of a supply system grounding conductor is also adequate for equipment grounding requirements, this conductor may be used for the combined purpose. Equipment referred to includes the frames and enclosures of supply system control and auxiliary components, conductor raceways, cable shields, and other enclosures.

94. Grounding Electrodes

The grounding electrode shall be permanent and adequate for the electrical system involved. A common electrode or electrode system shall be employed for grounding the electrical system and the conductor enclosures and equipment served by that system. This may be accomplished by interconnecting these elements at the "Point of Connection of Grounding Conductor," Rule 92.

Grounding electrodes shall be one of the following:

A. Existing Electrodes

Existing electrodes consist of conducting items installed for purposes other than grounding:

1. Metallic Water Piping System

   Extensive metallic underground cold water piping systems may be used as grounding electrodes.

   NOTE: Such systems normally have very low resistance to earth and have been extensively used in the past. They are the preferred electrode type where they are readily accessible.

   EXCEPTION: Water systems with nonmetallic, noncurrent-carrying pipe or insulating joints are not suitable for use as grounding electrodes.

2. Local Systems

   Isolated buried metallic cold water piping connecting to wells having sufficiently low measured resistance to earth may be used as grounding electrodes.

   NOTE: Care should be exercised to ensure that all parts that might become disconnected are effectively bonded together.

3. Steel Reinforcing Bars in Concrete Foundations and Footings

   The reinforcing bar system of a concrete foundation or footing which is not insulated from direct contact with earth, and which extends at least 3 feet (900 mm) below grade, constitutes an effective and acceptable type of grounding electrode. Where steel supported on this foundation is to be used as a grounding conductor (tower, structure, etc), it shall be interconnected by bonding between anchor bolts and reinforcing bars or by cable from the reinforcing bars to the structure above the concrete.

   The normally applied steel ties are considered to provide adequate bonding between bars of the reinforcing cage.

   NOTE: Where reinforcing bars in concrete are not suitably connected to a metal structure above the concrete, and the latter structure is subjected to grounding discharge currents (even connected to another electrode), there is likelihood of damage to the intervening concrete from ground-seeking current passing through the semiconducting concrete.

B. Made Electrodes

1. General

   Where made electrodes are used, they shall as far as practical penetrate permanent moisture level and below the frostline. Made electrodes shall be of metal or combinations of metals which do not corrode excessively under the existing conditions for the expected service life.

All outer surfaces of made electrodes shall be conductive, that is, not having paint, enamel, or other insulating type covering.

2. Driven Rods
   a. Driven rods may be sectional; the total length shall be not less than 8 ft (2.45 m). Iron or steel rods shall have a cross-sectional dimension of not less than 5/8 inch. Copper-clad, stainless steel or stainless steel-clad rods shall have a cross-sectional dimension of not less than 1/2 inch.
   b. Longer rods or multiple rods may be used to reduce the ground resistance. Spacing between multiple rods should be not less than 6 ft (1.8 m).
   c. Driven depth shall be not less than 8 ft (2.45 m). The upper end shall be flush with or below the ground level unless suitably protected.

      EXCEPTION 1: Where rock bottom is encountered, driven depth may be less than 8 ft (2.45 m), or other types of electrode employed.

      EXCEPTION 2: When contained within pad-mounted equipment or similar enclosures, the driven depth may be reduced to not less than 7.5 ft (2.3 m).
3. Buried Wire, Strips, or Plates

   In areas of high soil resistivity or shallow bedrock, or where lower resistance is required than attainable with driven rods, one or more of the following electrodes may be more useful:
   a. Wire

      Bare wires 0.162 inch in diameter or larger, conforming to Rule 93E5, buried in earth at a depth not less than 18 inches (450 mm) and not less than 100 feet (30 m) total in length, laid approximately straight, constitute an acceptable made electrode. (This is frequently designated a counterpoise.) The wire may be in a single length, or may be several lengths connected at ends or at some point away from the ends. The wire may take the form of a network with many parallel wires spaced in two-dimensional array, referred to as a grid.

EXCEPTION 1: Where rock bottom is encountered, burial depth may be less than 18 inches (450 mm).

EXCEPTION 2: Other lengths or configurations may be used if their suitability is supported by a qualified engineering study.

b. Strips

Strips of metal not less than 10 feet (3.0 m) in total length and with total (two sides) surface not less than 5 square feet (0.47 m^2) buried in soil at a depth not less than 18 inches (450 mm) constitute an acceptable made electrode. Ferrous metal electrodes shall be not less than 1/4 inch in thickness and nonferrous metal electrodes not less than 0.06 inches.

NOTE: Strip electrodes are frequently useful in rocky areas where only irregularly shaped pits are practical to excavate.

c. Plates or Sheets

Metal plates or sheets having not less than 2 square feet (0.185 m^2) of surface exposed to the soil, and at a depth of not less than 5 feet (1.5 m), constitute an acceptable made electrode. Ferrous metal electrodes shall be not less than 1/4 inch in thickness and nonferrous metal electrodes not less than 0.06 inches.

4. Pole Butt Plates and Wire Wraps

a. General

In areas of very low soil resistivity there are two constructions, described in specifications b and c below, which may provide effective grounding electrode functions although they are inadequate in most other locations. Where these have been proven to have adequately low earth resistance by the application of Rule 96, two such electrodes may be counted as one made electrode and ground for application of Rules 92C1a, 92C2b, 97C, and 96A3; however, these types shall not be the sole grounding electrode at transformer locations.

b. Pole Butt Plates

Subject to the limitations of Rule 94B4a, a pole butt plate on the base of a wooden pole, possibly folded up around the base of the pole butt, may be considered an acceptable electrode in locations where the limitations of Rule 96 are met. The plates shall be not less than 1/4 inch thick if of ferrous metal and not less than 0.06 inch thick if of nonferrous metal. Further, the plate area exposed to the soil shall be not less than 0.5 square feet (0.046 m^2).

c. Wire Wrap

Subject to the limitations of Rule 94B4a, made electrodes may be wire attached to the pole previous to the setting of the pole. The wire shall be of copper or other metals which will not corrode excessively under the existing conditions and shall have a continuous bare or exposed length below ground level of not less than 12 ft (3.7 m), shall extend to the bottom of the pole, and shall not be smaller than AWG No 6.

5. Concentric Neutral Cable

Systems employing extensive [100 ft (30 m) minimum length] buried bare concentric neutral cable in contact with the earth may employ the concentric neutral as a grounding electrode. The concentric neutral may be covered with a semi-conducting jacket which has a radial resistivity not exceeding 100 meter ohms and which will remain essentially stable in service. The radial resistivity of the jacket material is that value calculated from measurements on a unit length of cable, of the resistance between the concentric neutral and a surrounding conducting medium. Radial resistivity equals resistance of unit length times the surface area of jacket divided by the average thickness of the jacket over the neutral conductors. All dimensions are to be expressed in meters.

6. Concrete-Encased Electrodes

A metallic wire, rod, or structural shape, meeting Rule 93E5 and encased in concrete which is not insulated from direct contact with earth shall constitute an acceptable ground electrode. The concrete depth below grade shall not be less than 1 ft (300 mm), and a depth of 2 1/2 ft (750 mm) is recommended. Wire shall be no smaller than AWG No 4 if copper, or 3/8 inch diameter if steel. It shall be not less than 20 ft (6.1 m) long, and shall remain entirely within the concrete except for the external connection. The conductor should be run as straight as practical.

The metal elements may be composed of a number of shorter lengths arrayed within the concrete and connected together (for example, the reinforcing system in a structural footing).

EXCEPTION: Other wire length or configurations may be used if their suitability is supported by a qualified engineering study.

NOTE 1: The lowest resistance per unit wire length will result from a straight wire installation.

NOTE 2: The outline of the concrete need not be regular, but may conform to an irregular or rocky excavation.

NOTE 3: Concrete-encased electrodes are frequently more practical or effective than driven rods or strips or plates buried directly in earth.

95. Method of Connection to Electrode

A. Ground Connections

The ground connections shall be as accessible as practical and shall be made to the electrode by methods providing the required permanence and ampacity, such as:

1. A permanently effective clamp, fitting, braze, or weld.
2. A bronze plug which has been tightly screwed into the electrode.
3. For steel-framed structures, employing a concrete-encased reinforcing bar electrode, a steel rod similar to the reinforcing bar shall be used to join, by welding, a main vertical reinforcing bar to an anchor bolt. The bolt shall be substantially and permanently connected to the baseplate of the steel column supported on that footing. The electrical system may then be connected (for grounding) to the building frame by welding or by a bronze bolt tapped into a structural member of that frame.

4. For nonsteel frame structures employing a concrete-encased rod or wire electrode, an insulated copper conductor of size meeting the requirements of Rule 93C (except not smaller than AWG No 4) shall be connected to the steel rod or wire using a cable clamp suitable for steel cable. This clamp and all the bared portion of the copper conductor including ends of exposed stands within the concrete shall be completely covered with mastic or sealing compound before concrete is poured to minimize the possibility of galvanic corrosion. The copper conductor end shall be brought to or out of the concrete surface at the required location for connection to the electrical system. If the copper wire is carried beyond the surface of the concrete, it shall be no smaller than AWG No 2.

   Alternately, the copper wire may be brought out of the concrete at the bottom of the hole and carried external to the concrete for surface connection.

B. Point of Connection to Piping Systems

1. The point of connection of a grounding conductor to a metallic water piping system shall be as near as is practical to the water-service entrance to the building or near the equipment to be grounded and shall be accessible. If a water meter is between the point of connection and the underground water pipe, the metallic water piping system shall be made electrically continuous by bonding together all parts between the connection and the pipe entrance which may become disconnected, such as meters and service unions.
2. Made grounds or grounded structures should be separated by 10 ft (3.0 m) or more from pipelines used for the transmission of flammable liquids or gases operating at high pressure (150 pounds per square inch [1030 kPa] or greater) unless they are electrically interconnected and cathodically protected as a single unit. Grounds within 10 ft (3.0 m) of such pipelines should be avoided or shall be coordinated so that hazardous alternating current conditions will not exist and cathodic protection of the pipeline will not be nullified.

C. Contact Surfaces

If any coating of nonconducting material, such as enamel, rust, or scale, is present on electrode contact surfaces, at the point of connection, such a coating shall be thoroughly removed where required to obtain the requisite good connection. Special fittings so designed as to make such removal of nonconducting coatings unnecessary may also be used.

96. Ground Resistance

A. Requirements

The grounding electrode system may consist of one or more interconnected electrodes. It shall have a resistance to ground low enough to minimize hazards to personnel and to permit prompt operation of circuit protective devices.

1. Supply Stations

Where very high voltages and currents are involved, such as in large substations, extensive grounding grid systems of multiple buried wires and rods and other protective means may be required.

NOTE: It is recommended that the combination of maximum local ground fault current and impedance of the grounding system not exceed values that will limit exposure potentials to the following:

$$E_{step} = (1000 + 6\,\rho s)\,\frac{0.116}{\sqrt{t}}$$

$$E_{touch} = (1000 + 1.5\,\rho s)\,\frac{0.116}{\sqrt{t}}$$

where

E_{step} = maximum tolerable voltage difference between any two points on the ground surface which can be touched simultaneously by 2 (separated) feet

E_{touch} = maximum tolerable voltage difference between any point on the ground where a man may stand and any point which can be touched simultaneously by either hand

ρs = resistivity of the soil near the surface in ohm-meters (divide the ohm-centimeter value by 100 to obtain this)

t = time of exposure in seconds (clearing time of system overcurrent equipment)

2. Single-Grounded (Unigrounded or Delta) Systems
Individual made electrodes shall, where practical, have a resistance to ground not exceeding 25 ohms. If a single electrode resistance exceeds 25 ohms, two electrodes connected in parallel shall be used.

3. Multiple-Grounded Systems
The neutral, which shall be of sufficient size and ampacity for the duty involved, shall be connected to a made or existing electrode at each transformer location and at a sufficient number of additional points with made or existing electrodes to total not less than four grounds in each mile (1.6 km) of the entire line, not including grounds at individual services.
EXCEPTION: Where underwater crossings are encountered, the requirement of made electrodes to total not less than four grounds in each mile (1.6 km) of the entire line does not apply for the underwater portion if the neutral is of sufficient size and capacity for the duty involved and the requirements of Rule 92B2 are met.
NOTE: Multiple-grounding systems extending over a substantial distance are more dependent on the multiplicity of grounding electrodes than on the resistance to ground of any individual electrode. Therefore, no specific values are imposed for the resistance of individual electrodes.

97. Separation of Grounding Conductors

A. Except as permitted in Rule 97B, grounding conductors from equipment and circuits of each of the following classes shall be run separately to the grounding electrode for each of the following classes:

1. Surge arresters of circuits over 750 volts, and frames of any equipment operating at over 750 volts
2. Lighting and power circuits under 750 volts
3. Lightning rods, unless attached to a grounded metal supporting structure.

Alternatively, the grounding conductors shall be run separately to a sufficiently heavy ground bus or system ground cable which is well connected to ground at more than one place.

B. The grounding conductors of either of the equipment classes detailed in Rules 97A1 and 97A2 may be interconnected utilizing a single grounding conductor, provided:
   1. There is a direct earth grounding connection at each surge arrester location and
   2. The secondary neutral is common with, or connected to, a primary neutral meeting the grounding requirements of Rule 97C.

C. Primary and secondary circuits utilizing a single conductor as a common neutral shall have at least four ground connections on such conductor in each mile (1.6 km) of line, exclusive of ground connections at customers' service equipment.

D. Ungrounded or Single-Grounded Systems and Multiple-Grounded Systems
   1. Ungrounded or Single-Grounded Systems
      Where the secondary neutral is not interconnected with the primary surge arrester grounding conductor as in Rule 97B, interconnection may be made through a spark gap or device which performs an equivalent function. The gap or device shall have a 60 Hz breakdown voltage of at least twice the primary circuit voltage but not necessarily more than 10 kV. At least one other grounding connection on the secondary neutral shall be provided with its grounding electrode located at a distance of not less than 20 ft (6.1 m) from the surge arrester grounding electrode in addition to customers' grounds at each service entrance.
   2. Multiple-Grounded Systems
      On multiple-grounded systems the primary and secondary neutrals should be interconnected according to Rule 97B. However, where it is necessary to separate the neutrals, interconnection of the neutrals shall be made through a spark gap or a device which performs an equivalent function. The gap or device shall have a 60 Hz breakdown voltage not exceeding 3 kV. At least one other grounding connection on the secondary neutral shall be provided with its grounding electrode located at a distance not less than 6 ft (1.80 m) from the primary neutral and surge arrester grounding electrode in addition to the customers' grounds at each service entrance. Since a dif-

ferent potential can exist where primary and secondary neutrals are not directly interconnected, the secondary grounding conductor shall be insulated for 600 V.

NOTE: Cooperation of all communication and supply utilities, customers of these utilities, and others may be necessary to obtain effective isolation between primary and secondary neutrals.

E. Where separate electrodes are used for system isolation, separate grounding conductors shall be used. Where multiple electrodes are used to reduce grounding resistance, they may be bonded together and connected to a single grounding conductor.

F. Made electrodes used for grounding surge arresters of ungrounded supply systems operated at potentials exceeding 15 kilovolts phase to phase should be located at least 20 feet (6.1 m) from buried communications cables. Where lines with lesser separations are to be constructed, reasonable advance notice should be given to the owners or operators of the affected systems.

98. Number 98 not used in this edition.

99. Grounding Methods for Telephone and Other Communication Apparatus on Circuits Exposed to Supply Lines or Lightning

Protectors and, where required, exposed noncurrent-carrying metal parts located in central offices or outside installations shall be grounded in the following manner:

A. Electrode

The grounding conductor shall be connected to an acceptable grounding electrode as follows:

1. Where available and where the supply service is grounded to an acceptable electrode as described in Rule 94, to the grounded metallic supply service conduit, service equipment enclosure, grounding electrode conductors, or grounding electrode conductors' metal enclosure.
2. Where the grounding means of Rule 99A1 is not available, to a grounding electrode as described in Rule 94A.

3. Where the grounding means of Rule 99A1 or 99A2 are not available, to a grounding electrode as described in Rule 94B.

   EXCEPTION: A variance to Rule 94B2 is allowed for this application. Iron or steel rods may have a cross-sectional dimension of not less than 1/2 inch and a length of not less than 5 ft (1.50 m). The driven depth shall be 5 ft (1.50 m), subject to the exception of Rule 94B2.

B. Electrode Connection

The grounding conductor shall preferably be made of copper (or other material which will not corrode excessively under the prevailing conditions of use) and shall be not less than AWG No 14 (0.064 inch) in size. The grounding conductor shall be attached to the electrode by means of a bolted clamp or other suitable methods.

C. Bonding of Electrodes

A bond not smaller than AWG No 6 (0.162 inch) copper or equivalent shall be placed between the communication grounding electrode and the supply system neutral grounding electrode where separate electrodes are used in or on the same building or structure.

PART 1. Rules for the Installation and Maintenance of Electric Supply Stations and Equipment

Section 10. Purpose and Scope of Rules

100. Purpose

The purpose of Part 1 of this code is the practical safeguarding of persons during the installation, operation, or maintenance of electric supply stations and their associated equipment.

101. Scope

Part 1 of this code covers the electric supply conductors and equipment, along with the associated structural arrangements in electric supply stations, which are accessible only to qualified personnel. It also covers the conductors and equipment employed primarily for the utilization of electric power when such conductors and equipment are used by the utility in the exercise of its function as a utility.

Section 11. Protective Arrangements in Electric Supply Stations

110. General Requirements

A. Enclosure of Equipment

Rooms and spaces in which electric supply conductors or equipment are installed shall be so arranged with fences, screens, partitions or walls as to minimize the possibility of entrance of unauthorized persons or interference by them with equipment inside. Entrances not under observation of an authorized attendant shall be kept locked.

Warning signs shall be displayed at entrances.

Metal fences when used to enclose electric supply stations having energized electrical conductors or equipment shall have a height not less than 7 ft (2.13 m) overall and shall be grounded in accordance with Section 9.

The requirements for fence height may be satisfied with any one of the following:

1. Fence fabric, 7 ft (2.13 m) or more in height.
2. A combination of 6 ft (1.80 m) or more of fence fabric and a 1 ft (300 mm) extension utilizing three or more strands of barbed wire.
3. Other types of construction, such as nonmetallic material, which present equivalent barriers to climbing or other unauthorized entry.

NOTE: ANSI/IEEE Std 1119-1988, IEEE Guide for Fence Safety Clearances in Electric Supply Stations [29], may be utilized to provide guidance for station fence safety clearances.

B. Rooms and Spaces

All rooms and spaces in which electric supply equipment is installed shall comply with the following requirements:

1. Construction

   They shall be as much as practical noncombustible.
2. Use

   They should be as much as practical free from combustible materials, dust, and fumes and shall not be used for manufacturing or for storage, except for minor parts essential to the maintenance of the installed equipment. (For battery areas, see Section 14; for auxiliary equipment in classified locations, see Rule 127.)
3. Ventilation

   There should be sufficient ventilation to maintain operating temperatures within ratings, arranged to minimize accumulation of airborne contaminants under any operating conditions.
4. Moisture and Weather

   They should be dry. In outdoor stations or stations in wet tunnels, subways or other moist or high humidity locations, the equipment shall be suitably designed to withstand the prevailing atmospheric conditions.

C. Electric Equipment

All stationary equipment shall be supported and secured in a manner consistent with reasonably expected conditions of service. Consideration shall be given to the fact that certain heavy equipment, such as transformers, can be secured in place by their weight. However, equipment that generates dynamic forces during operation may require appropriate additional measures.

111. Illumination

A. Under Normal Conditions

Rooms and spaces shall have means for artificial illumination. Illumination levels not less than those listed in Table 111-1 are recommended for safety to be maintained on the task.

B. Emergency Lighting

1. A separate emergency source of illumination with automatic initiation, from an independent generator, storage battery or other suitable source, shall be provided in every attended station.
2. Emergency lighting of 1 footcandle (11 lux) shall be provided in exit paths from all areas of attended stations. Consideration must be given to the type of service to be rendered whether of short time or long duration. The minimum duration shall be 1 1/2 hours. It is recommended that emergency circuit wiring shall be kept independent of all other wiring and equipment.

C. Fixtures

Arrangements for permanent fixtures and plug receptacles shall be such that portable cords need not be brought into dangerous proximity to live or moving parts. All lighting shall be controlled and serviced from safely accessible locations.

D. Attachment Plugs and Receptacles for General Use

Portable conductors shall be attached to fixed wiring only through separable attachment plugs which will disconnect all poles by one operation. Receptacles installed on two or three wire single phase, ac branch circuits shall be of the grounding type. Receptacles connected to circuits having different voltages, frequencies or types of current (ac or dc) on the same premises shall be of such design that attachment plugs used on such circuits are not interchangeable.

E. Receptacles in Damp or Wet Locations

All 120 V ac permanent receptacles shall either be provided with ground fault interrupter (GFI) protection, or be on a grounded circuit which is tested at such intervals as experience has shown to be necessary.

Table 111-1 Illumination Levels

Location	(footcandles)	(lux)
Central Station		
Air conditioning equipment, air preheater and fan floor, ash sluicing	5	55
Auxiliaries, battery areas, boiler feed pumps, tanks, compressors, gage area	10	110
Boiler platforms	5	55
Burner platforms	10	110
Cable room, circulator, or pump bay	5	55
Chemical laboratory	25	270
Coal conveyor, crusher, feeder, scale areas, pulverizer, fan area, transfer tower	5	55
Condensers, deaerator floor, evaporator floor, heater floors	5	55
Control rooms		
Vertical face of switchboards		
Simples or section of duplex operator:		
Type A—Large centralized control room 66 inches above floor	25	270
Type B—Ordinary control room 66 inches above floor	15	160
Section of duplex facing away from operator	15	160
Bench boards (horizontal level)	25	270
Area inside duplex switchboards	5	55
Rear of all switchboard panels (vertical)	5	55
Dispatch boards		
Horizontal plane (desk level)	25	270
Vertical face of board (48 inches above floor, facing operator):		
System load dispatch room	25	270
Secondary dispatch room	15	160
Hydrogen and carbon dioxide manifold area	10	110
Precipitators	5	55
Screen house	10	110

(continued on next page)

Table 111-1 *(Continued)*

Location	(footcandles)	(lux)
Soot or slag blower platform	5	55
Steam headers and throttles	5	55
Switchgear, power	10	110
Telephone equipment room	10	110
Tunnels or galleries, piping	5	55
Turbine bay sub-basement	10	110
Turbine room	15	160
Visitors' gallery	10	110
Water treating area	10	110
Central Station (Exterior)		
Catwalks	2	22
Cinder dumps	0.2	2.2
Coal storage area	0.2	2.2
Coal unloading		
Dock (loading or unloading zone)	5	55
Barge storage area	0.5	5.5
Car dumper	0.5	5.5
Tipple	5	55
Conveyors	2	22
Entrances		
Generating or service building		
Main	10	110
Secondary	2	22
Gate house		
Pedestrian entrance	10	110
Conveyor entrance	5	55
Fence	0.2	2.2
Fuel-oil delivery headers	5	55
Oil storage tanks	1	11
Open yard	0.2	2.2
Platforms—boiler, turbine deck	5	55
Roadway		
Between or along buildings	1	11
Not bordered by buildings	0.5	5.5
Substation		
General horizontal	2	22
Specific vertical (on disconnects)	2	22

112. Floor, Floor Openings, Passageways, Stairs

A. Floors

Floors shall have even surfaces and afford secure footing. Slippery floors or stairs should be provided with antislip covering.

B. Passageways

Passageways, including stairways, shall be unobstructed and shall, where practical, provide at least 7 ft (2.13 m) head room. Where the preceding requirements are not practical, the obstructions should be painted, marked or indicated by warning signs and the area properly lighted.

C. Railings

All floor openings without gratings or other adequate cover and raised platforms and walkways in excess of 1 ft (300 mm) in height shall be provided with railings. Openings in railings for units such as fixed ladders, cranes, and the like shall be provided with adequate guards such as grates, chains, or sliding pipe sections.

D. Stair Guards

All stairways consisting of four or more risers shall be provided with handrails.

NOTE: For additional information see ANSI A12.1-1973 [1].

E. Top Rails

All top rails shall be kept unobstructed for a distance of 3 in (75 mm) in all directions except from below at supports.

113. Exits

A. Clear Exits

Each room or space and each working space about equipment shall have a means of exit which shall be kept clear of all obstructions.

B. Double Exits

If the plan of the room or space and the character and arrangement of equipment are such that an accident would be likely to close or make inaccessible a single exit, a second exit shall be provided.

C. Exit Doors

Exit doors shall swing out and be equipped with panic bars, pressure plates, or other devices that are normally latched but open under simple pressure.

EXCEPTION: This rule does not apply to exit doors in buildings and rooms containing low-voltage, nonexplosive equipment, and to gates in fences for outdoor equipment installations.

114. Fire Extinguishing Equipment

Fire extinguishing equipment approved for the intended use shall be conveniently located and conspicuously marked.

Section 12. Installation and Maintenance of Equipment

120. General Requirements

All electric equipment shall be constructed, installed, and maintained so as to safeguard personnel as far as practical.

121. Inspections

A. In-Service Equipment

Electric equipment shall be inspected and maintained at such intervals as experience has shown to be necessary. Equipment or wiring found to be defective shall be put in good order or permanently disconnected.

B. Idle Equipment

Infrequently used equipment or wiring shall be inspected and tested before use to determine its fitness for service. Idle equipment energized but not connected to a load shall be inspected and maintained at such intervals as experience has shown to be necessary.

C. Emergency Equipment

Equipment and wiring maintained for emergency service shall be inspected and tested at such intervals as experience has shown to be necessary to determine its fitness for service.

D. New Equipment

New equipment shall be inspected and tested before being placed in service.

EXCEPTION: The equipment to be tested does not include fittings, devices, appliances, fixtures or other hardware.

122. Guarding Shaft Ends, Pulleys, Belts, and Suddenly Moving Parts

A. Mechanical Transmission Machinery

The methods for safeguarding pulleys, belts and other equipment used in the mechanical transmission of power shall be in accordance with ANSI/ASME B15.1-1984 [23].

B. Suddenly Moving Parts

Parts of equipment which move suddenly in such a way that persons in the vicinity are likely to be injured by such movement shall be guarded or isolated.

123. Protective Grounding

A. Protective Grounding or Physical Isolation of Noncurrent-Carrying Metal Parts

All electric equipment shall have the exposed noncurrent-carrying metal parts, such as frames of generators and switchboards, cases of transformers, switches and operating levers effectively grounded or physically isolated. All metallic guards including rails, screen fences, etc, about electric equipment shall be effectively grounded.

B. Grounding Method

All grounding which is intended to be a permanent and effective protective measure, such as surge arrester grounding, grounding of circuits, equipment, or wire raceways, shall be made in accordance with the methods specified in Section 9 of this code.

NOTE: For additional information see ANSI/IEEE Std 80-1986 [26].

C. Provision for Grounding Equipment During Maintenance

Electric equipment or conductors normally operating at more than 600 V between conductors, on or about which work is occasionally done while isolated from a source of electric energy by disconnecting or isolating switches only, shall be provided with some means for grounding, such as switches, connectors or a readily accessible means for connecting a portable grounding conductor. When necessary, grounding may be omitted on conductors normally operating at 25 kV or less and not influenced by higher voltage conductors, where visible openings in the source of supply are available, and are properly tagged in the open position. (See Part 4 of this code.)

124. Guarding Live Parts

A. Where Required

1. Guards shall be provided around all live parts operating above 150 V to ground without an adequate insulating covering, unless their location gives sufficient horizontal or vertical or a combination of these clearances to minimize the possibility of accidental human contact. Clearances from live parts to any permanent supporting surface for workers shall equal or exceed either of those shown in Table 124-1 and illustrated in Fig 124-1.
2. Parts over or near passageways through which material may be carried, or in or near spaces such as corridors, storerooms and boiler rooms used for nonelectrical work shall be guarded or given clearances in excess of those specified such as may be necessary to secure reasonable safety. The guards shall be substantial and completely shield or enclose the live parts without openings. In spaces used for nonelectrical work, guards should be removable only by means of tools or keys.
3. Each portion of parts of indeterminate potential, such as telephone wires exposed to induction from high voltage lines, ungrounded neutral connections, ungrounded frames, ungrounded parts of insulators or surge arresters, or ungrounded instrument cases connected directly to a high voltage circuit, shall be guarded in accordance with Rule 124A1 on the basis of the maximum voltage which may be present on the surface of that portion. The vertical clearance above grade of the bottom of such part shall be not less than 8 ft 6 in (2.60 m) unless it is enclosed or guarded in accordance with Rule 124C.

B. Strength of Guards

Guards shall be sufficiently strong and shall be supported rigidly and securely enough to prevent them from being displaced or dangerously deflected by a person slipping or falling against them.

C. Types of Guards

1. Location or Physical Isolation

   Parts having clearances equal to or greater than specified in Table 124-1 are guarded by location. Parts are guarded by isolation when all entrances to enclosed spaces, runways, fixed ladders, and the like are kept locked, barricaded, or roped off and warning signs are posted at all entrances.

2. Shields or Enclosures

   Guards less than 4 inches (100 mm) outside of the guard zone shall completely enclose the parts from contact up to the heights listed in column 2 of Table 124-1. They shall not be closer to the live parts than listed in column 4 of Table 124-1, except when suitable insulating material is used with circuits of less than 2500 V to ground. (See NOTE under Table 124-1). If more than 4 inches (100 mm) outside the guard zone, the guards shall extend at least 8 ft 6 in (2.60 m) above the floor. Covers or guards, which must at any time be removed while the parts they guard are live shall be arranged so that they cannot readily be brought into contact with live parts.

3. Railings

   Railings are not substitutes for complete guards. If the vertical distance in Table 124-1 cannot be obtained, railings may be used. Railings, if used, shall be located at a horizontal distance of at least 3 ft (900 mm) [and preferably not more than 4 ft (1.20 m)] from the nearest point of the guard zone which is less than 8 ft, 6 in (2.60 m) above the floor (see Fig 124-2).

4. Mats

   Mats of rubber or other suitable insulating material complying with ASTM D178-81 [37] may be used at switchboards, switches, or rotating machinery as supplementary protection.

5. Live Parts Below Supporting Surfaces for Persons

   The supporting surfaces for persons above live parts shall be without openings. Toe boards at least 6 inches (150 mm) high and handrails shall be provided at all edges.

6. Insulating Covering on Conductors or Parts

   Conductors and parts may be considered as guarded by insulation if they have either of the following:

a. Insulation covering of a type and thickness suitable for the voltage and conditions under which they are expected to be operated and if operating above 2500 V to ground having metallic insulation shielding or semiconducting shield in combination with suitable metallic drainage which is grounded to an effective ground.

EXCEPTION: Nonshielded insulated conductors listed by a qualified testing laboratory shall be permitted for use up to 8000 V (phase-to-phase) when the conductors meet the requirements of ANSI/NFPA 70-1984 [31], Article 310-6.

b. Barriers or enclosures which are electrically and mechanically suitable for the conditions under which they are expected to be operated.

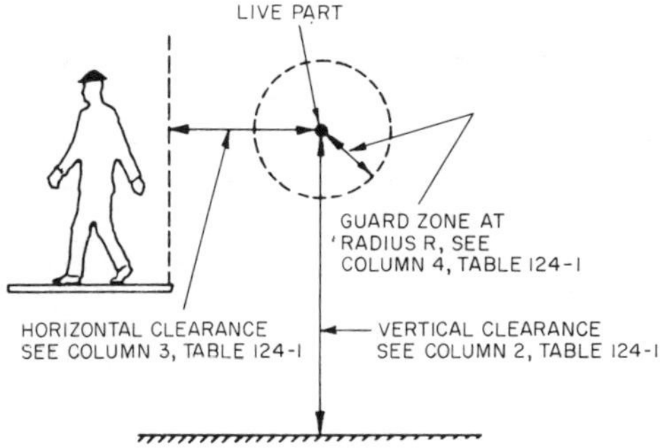

Fig 124-1
Clearance From Live Parts

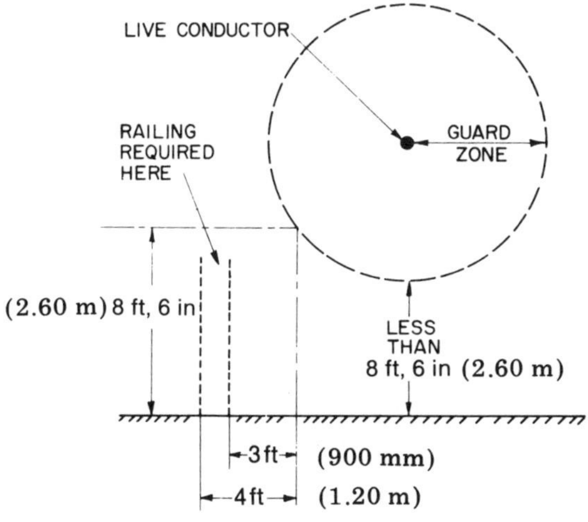

Fig 124-2
Railing Used As Guards

Table 124-1 Clearance from Live Parts **FT**

PART A — Low, Medium and High Voltages

Nominal voltage between phases (1)	Vertical clearance of unguarded parts (2)①		Horizontal clearance of unguarded parts (3)①		Clearance guard to live parts (4)①	
	Feet	Inches	Feet	Inches	Feet	Inches
151 to 600	8	8	3	4		2
2 400	8	9	3	4		3
7 200	8	10	3	4		4
13 800	9	0	3	6		6
23 000	9	3	3	9		9
34 500	9	6	4	0	1	0
46 000	9	10	4	4	1	4
69 000	10	5	4	11	1	11
115 000	11	7	6	1	3	1
138 000	12	2	6	8	3	8
161 000	12	10	7	4	4	4
230 000	14	10	9	4	6	4

(continued on next page)

Table 124-1 *(Continued)* **FT**

PART B — Extra high voltages (based on switching surge factors)②

Maximum design voltage between phases (1)	Switching surge factor③ per unit (A)④	Switching surge line to ground (B)④	Vertical clearance of unguarded parts (2)①		Horizontal clearance of unguarded parts (3)①		Clearance guard to live parts (4)①	
		kV	Ft	In	Ft	In	Ft	In
362 000	2.2 or below	650	15	6	10	0	7	0
	2.3	680	16	0	10	6	7	6
	2.4	709	16	6	11	0	8	0
	2.5	739	17	2	11	8	8	8
	2.6	768	17	9	12	3	9	3
	2.7	798	18	4	12	10	9	10
	2.8	828	18	11	13	5	10	5
	2.9	857	19	7	14	1	11	1
	3.0	887	20	2	14	8	11	8
550 000	1.8 or below	808	18	10	13	4	10	4
	1.9	853	19	6	14	0	11	0
	2.0	898	20	6	15	0	12	0
	2.1	943	21	6	16	0	13	0
	2.2	988	22	6	17	0	14	0
	2.3	1033	23	7	18	1	15	1
	2.4	1078	24	8	19	2	16	2
	2.5	1123	25	10	20	4	17	4
	2.6	1167	27	0	21	6	18	6
	2.7	1212	28	4	22	10	19	10
800 000	1.5	980	22	4	16	10	13	10
	1.6	1045	23	11	18	5	15	5
	1.7	1110	25	6	20	0	17	1
	1.8	1176	27	3	21	9	18	9
	1.9	1241	29	0	23	6	20	6
	2.0	1306	30	10	25	4	22	4
	2.1	1372	32	9	27	3	24	3
	2.2	1437	34	8	29	3	26	2
	2.3	1502	36	9	31	3	28	3
	2.4	1567	38	9	33	3	30	3

(continued on next page)

Table 124-1 *(Continued)* **FT**

PART C — Extra high voltages (based on BIL factors)②

Maximum design voltage between phases (1)	Basic impulse insulation level⑤ (BIL) (C)④	Vertical clearance of unguarded parts (2)①		Horizontal clearance of unguarded parts (3)①		Clearance guard to live parts (4)①	
	kV	ft	in	ft	in	ft	in
362 000	1050	15	6	10	0	7	0
362 000	1300	17	2	11	8	8	8
550 000	1550	18	10	13	4	10	4
550 000	1800	20	6	15	0	12	0
800 000	2050	22	5	16	11	13	11

Notes and explanations to terms used in Table 124:

① Interpolate for intermediate values. The clearances in column 4 of this table are solely for guidance in installing guards without definite engineering design and are not to be considered as a requirement for such engineering design. For example, the clearances in the tables above are not intended to refer to the clearances between live parts and the walls of the cells, compartments or similar enclosing structures. They do not apply to the clearances between bus bars and supporting structures nor to clearances between the blade of a disconnecting switch and its base. However, where surge protective devices are applied to protect the live parts, the vertical clearances, Column 2 of Table 124-1 Part A may be reduced provided the clearance is not less than eight feet and six inches plus the electrical clearance between energized parts and ground as limited by the surge protective devices.

② Clearances shall satisfy either switching surge or BIL duty requirements, whichever are greater.

③ Switching Surge Factor—an expression of the maximum Switching Surge Crest Voltage in terms of the maximum operating Line to Neutral Crest Voltage of the power system.

④ The values of columns A, B, and C are power system design factors that shall correlate with selected clearances. Adequate data to support these design factors should be available.

⑤ The selection of station BIL shall be coordinated with surge protective devices when using BIL to determine clearance. BIL—Basic Impulse Insulation Level— For definition and application see ANSI C92.1-1982 [16].

(continued on next page)

Table 124-1 Clearance from Live Parts **M**

PART A—Low, Medium and High Voltages

Nominal voltage between phases (1)	Vertical clearance of unguarded parts (2)①	Horizontal clearance of unguarded parts (3)①	Clearance guard to live parts (4)①
	m	m	mm
151 to 600	2.60	1.00	50
2 400	2.70	1.00	75
7 200	2.70	1.00	100
13 800	2.70	1.07	150
23 000	2.80	1.14	230
34 500	2.90	1.20	300
46 000	3.0	1.32	410
69 000	3.2	1.50	580
115 000	3.5	1.85	940
138 000	3.7	2.00	1.12 m
161 000	3.9	2.25	1.32 m
230 000	4.5	2.80	1.90 m

PART B—Extra high voltages (based on switching surge factors)② **M**

Maximum design voltage between phases (1)	Switching surge factor per unit (A) ④	Switching surge line to ground (B) ④	Vertical clearance of unguarded parts (2) ①	Horizontal clearance of unguarded parts (3) ①	Clearance guard to live parts (4) ①
		kV	m	m	m
362 000	2.2 or below	650	4.7	3.0	2.13
	2.3	680	4.9	3.2	2.30
	2.4	709	5.0	3.4	2.45
	2.5	739	5.2	3.6	2.60
	2.6	768	5.4	3.7	2.80
	2.7	798	5.6	3.9	3.0
	2.8	828	5.8	4.1	3.2
	2.9	857	6.0	4.3	3.4
	3.0	887	6.1	4.5	3.6

(continued on next page)

Table 124-1 *(Continued)*

PART B—Extra high voltage (based on switching surge factors) ② **M**

Maximum design voltage between phases (1)	Switching surge factor per unit (A) ④	Switching surge line to ground (B) ④	Vertical clearance of unguarded parts (2) ①	Horizontal clearance of unguarded parts (3) ①	Clearance guard to live parts (4) ①
550 000	1.8 or below	808	5.7	4.1	3.2
	1.9	853	5.9	4.3	3.4
	2.0	898	6.2	4.6	3.6
	2.1	943	6.6	4.9	4.0
	2.2	988	6.9	5.2	4.3
	2.3	1033	7.2	5.5	4.6
	2.4	1078	7.5	5.8	4.9
	2.5	1123	7.9	6.2	5.3
	2.6	1167	8.2	6.6	5.6
	2.7	1212	8.6	7.0	6.0
800 000	1.5	980	6.8	5.1	4.2
	1.6	1045	7.3	5.6	4.7
	1.7	1110	7.8	6.1	5.2
	1.8	1176	8.3	6.6	5.7
	1.9	1241	8.8	7.2	6.2
	2.0	1306	9.4	7.7	6.8
	2.1	1372	10.0	8.3	7.4
	2.2	1437	10.6	8.9	8.0
	2.3	1502	11.2	9.5	8.6
	2.4	1567	11.8	10.0	9.2

(continued on next page)

Table 124-1 *(Continued)*

PART C—Extra High Voltage (based on BIL factors)② **M**

Maximum design voltage between phases (1)	Basic impulse insulation⑤ level (BIL) (C)④	Vertical clearance of unguarded parts (2) ①	Horizontal clearance of unguarded parts (3) ①	Clearance guard to live parts (4) ①
	kV	m	m	m
362 000	1050	4.7	3.0	2.13
362 000	1300	5.2	3.6	2.60
550 000	1550	5.7	4.1	3.2
550 000	1800	6.2	4.6	3.6
800 000	2050	6.8	5.2	4.2

Notes and explanations to terms used in Table 124:

① Interpolate for intermediate values. The clearances in column 4 of this table are solely for guidance in installing guards without definite engineering design and are not to be considered as a requirement for such engineering design. For example, the clearances in the tables above are not intended to refer to the clearances between live parts and the walls of the cells, compartments or similar enclosing structures. They do not apply to the clearances between bus bars and supporting structures nor to clearances between the blade of a disconnecting switch and its base. However, where surge protective devices are applied to protect the live parts, the vertical clearances, Column 2 of Table 124-1 Part A may be reduced provided the clearance is not less than 2.6 m plus the electrical clearance between energized parts and ground as limited by the surge protective devices.

② Clearances shall satisfy either switching surge or BIL duty requirements, whichever are greater.

③ Switching Surge Factor—an expression of the maximum Switching Surge Crest Voltage in terms of the maximum operating Line to Neutral Crest Voltage of the power system.

④ The values of columns A, B, and C are power system design factors that shall correlate with selected clearances. Adequate data to support these design factors should be available.

⑤ The selection of station BIL shall be coordinated with surge protective devices when using BIL to determine clearance. BIL—Basic Impulse Insulation Level— For definition and application see ANSI C92.1-1982 [16].

125. Working Space About Electric Equipment

A. Working Space (600 Volts or Less)

Access and working space shall be provided and maintained about electric equipment to permit ready and safe operation and maintenance of such equipment.

1. Clear Spaces

   Working space required by this section shall not be used for storage. When normally enclosed energized parts are exposed for inspection or servicing, the working space, if in a passageway or general open space, shall be guarded.

2. Access and Entrance to Working Space

   At least one entrance shall be provided to give access to the working space about electrical equipment.

3. Working Space

   The working space in the direction of access to energized parts operating at 600 volts or less which require examination, adjustment, servicing, or maintenance while energized shall be not less than indicated in Table 125-1. In addition to the dimensions shown in Table 125-1, the working space shall be not less than 30 in (750 mm) wide in front of the electric equipment. Distances shall be measured from the energized parts if such are exposed or from the enclosure front or opening if such are enclosed. Concrete, brick, or tile walls shall be considered grounded.

4. Headroom Working Space

   The headroom of working spaces about switchboards or control centers shall be not less than 7 ft (2.13 m).

5. Front Working Space

   In all cases where there are energized parts normally exposed on the front of switchboards or motor control centers, the working space in front of such equipment shall not be less than 3 ft (900 mm).

B. Working Space Over 600 Volts

Working space shall be in accordance with Table 124-1 clearances for guarding.

Table 125-1
Working Space

Voltage to ground	Clear Distance (ft) Condition: 1	(mm) 1	(ft) 2	(mm) 2	(ft) 3	(mm) 3
0–150	3	900	3	900	3	900
151–600	3	900	31/2	1.07 m	4	1.20 m

Where the *Conditions* are as follows:

1. Exposed energized parts on one side and no energized or grounded parts on the other side of the working space, or exposed energized parts on both sides effectively guarded by suitable wood or other insulating materials. Insulated wire or insulated busbars operating at not over 300 V shall not be considered energized parts.
2. Exposed energized parts on one side and grounded parts on the other side.
3. Exposed energized parts on both sides of the work space [not guarded as provided in Condition 1] with the operator between.

EXCEPTION: Working space shall not be required in back of assemblies, such as dead-front switchboards or motor control centers where there are no renewable or adjustable parts such as fuses or switches on the back and where all connections are accessible from locations other than the back.

126. Equipment for Work on Energized Parts

When it is necessary for personnel to move themselves, material, or tools within the guard zone of unguarded energized parts, protective equipment shall be provided.

This protective equipment shall be periodically inspected, tested, and kept in a safe condition. Protective equipment shall be rated for not less than the voltage involved. Refer to Section 3 for a list of specifications for equipment.

127. Classified Locations

Electrical installations in classified areas shall meet the requirements of ANSI/NFPA 70-1987 [32], Articles 500 through 503 and Articles 511 through 517.

Specific classified areas in a power plant or substation and their classifications are identified in the following sections.

A. Coal-Handling Areas

1. Unventilated tunnels below stockpiles or surge piles and spaces inside, above or below coal storage silos or bunkers or other enclosed coal storage spaces where methane or coal dust may accumulate, are Class I, Division 1, Group D, and Class II, Division 1, Group G

locations except areas where coal dust resistivity is less than 10^5 ohm-cm which are Class II, Division 1, Group F locations.

2. Enclosed areas of preparation plants or coalhandling facilities where coal dust might accumulate are Class II, Division 1, Group G locations except areas where coal dust resistivity is less than 10^5 ohm-cm are Class II, Division 1, Group E locations.
3. Electrical equipment in other locations in which hazardous concentrations of flammable gases or vapors may exist continually, intermittently or periodically under normal operating conditions shall be in accordance with ANSI/NFPA 70-1987 [32], Articles 501, or be adequately ventilated.
4. The minimum acceptable requirements for adequate ventilation (pressurization) to reduce the classification of an enclosed area or enclosure within a Class I, Division 1 area to non-classified are:
   a. The ventilation system shall maintain at least 0.1 inch (2.5 mm) of positive water pressure in the area with all openings closed.
   b. The ventilation system shall provide a minimum velocity of 60 ft per min outward through each opening with all openings open at the same time.
   c. The ventilation system shall be interlocked so that on failure of the ventilation system, all power to the area shall be de-energized except to those devices which meet the Class I, Division 1 requirements without the ventilation system.
   d. The maximum operating temperature of any internal surface shall not exceed 80% of the ignition temperature (°C) of the hazardous material involved.
5. Locations in which combustible dust is or may be in suspension in the air continuously, intermittently, or periodically under normal operating conditions, or in quantities sufficient to produce explosions or ignitable mixtures, are classified as Class II, Division 1, Group G locations except areas where coal dust resistivity is less

than 10^5 ohm-cm are Class II, Division 1, Group E locations and all electrical equipment shall be installed and maintained in accordance with the requirements of ANSI/NFPA 70-1987 [32], Article 502.

6. Locations where dangerous concentrations of suspended dust are prevented during normal operation but where dust accumulations on electrical equipment may be sufficient to interfere with the safe dissipation of heat from electrical equipment or might be ignited by arcs, sparks, or burning material from such equipment are Class II, Division 2, Group G locations except areas where coal dust resistivity is less than 10^5 ohm-cm are Class II, Division 1, Group E locations and all electrical equipment shall be installed and maintained in accordance with the requirements of ANSI/NFPA 70-1987 [32], Article 502.
7. Enclosed sections where only wet coal is handled or enclosed sections so cut off as to be free from dangerous amounts of coal dust are not classified. Coal shall be considered to be wet if enough water sprays are installed and maintained to prevent more than 0.3 ounce (8.5 g) of coal dust per cu ft (0.028 cu m) of enclosed air volume from being thrown into suspension or from accumulating on or in electrical equipment.
8. Locations having completely dust-tight pulverized fuel systems designed and installed in compliance with ANSI/NFPA 85F-1982 [33], shall not be considered classified.
9. Portable lamps for use in Fuel Bunkers or Bins shall be suitable for Class II, Division 1 locations.
10. Sparking electric tools shall not be used where flammable dust or dust clouds are present.
11. An equipment grounding conductor shall be carried with the power conductors and serve to ground the frames of all equipment supplied from that circuit. The origin of the grounding conductor shall be:
    a. Ungrounded Delta or Wye-Transformer frame ground.
    b. Grounded Delta or Wye-Transformer grounded secondary connection.

c. Resistance Grounded Wye—The grounded side of the grounding resistor.

12. Ungrounded systems should be equipped with a ground fault indicating device to give both a visual and audible alarm upon the occurrence of a ground fault in the system.

B. Flammable and Combustible Liquids

1. Flammable liquid shall mean a liquid having a flash point below 100 °F (38 °C) and having a vapor pressure not exceeding 40 pounds per sq in (275 kPa) (absolute) at 100 °F (38 °C) and shall be known as a Class I liquid.
2. Combustible liquid shall mean a liquid having a flash point greater than or equal to 100 °F (38 °C) and having a vapor pressure not exceeding 40 pounds per sq in (275 kPa) (absolute) at 100 °F (38 °C).
3. Class I liquids are subdivided as follows:
   a. Class IA includes those having flash points below 73 °F (23 °C) and having a boiling point below 100 °F (38 °C).
   b. Class IB includes those having flash points below 73 °F (23 °C).
   c. Class IC includes those having flash points at or above 73 °F (23 °C) and below 100 °F (38 °C).
4. Combustible liquids are subdivided as follows:
   a. Class II includes those having flash points equal to or greater than 100 °F (38 °C) but less than 140 °F (60 °C).
   b. Class IIIA includes those having flash points equal to or greater than 140 °F (60 °C) but less than 200 °F (93 °C).
   c. Class IIIB includes those having flash points greater than or equal to 200 °F (93 °C).

C. Flammable Liquid Storage Area

1. Electrical wiring and equipment located in inside storage rooms used for Class I liquids shall be approved for Class I, Division 2 locations (see Table 127-1).

D. Loading and Unloading Facilities

Electrical equipment located in the area shall comply with the requirements of Table 127-2.

1. Static Protection
   Bonding facilities for protection against static sparks during the loading of tank vehicles through open domes shall be provided (a) where Class I liquids are loaded, or (b) where Class II or Class III liquids are loaded into vehicles which may contain vapors from previous cargoes of Class I liquids.
   a. Protection as required in Rule 127D1 shall consist of a metallic bond wire permanently electrically connected to the fill stem or to some part of the rack structure in electrical contact with the fill stem. The free end of such wire shall be provided with a clamp or equivalent device for convenient attachment to some metallic part in electrical contact with the cargo tank of the tank vehicle.
   b. Such bonding connection shall be made fast to the vehicle or tank before dome covers are raised and shall remain in place until filling is completed and all dome covers have been closed and secured.

   EXCEPTION: Bonding as specified in Rules 127D1, 127D1a, 127D1b is not required:
   (1) Where vehicles are loaded exclusively with products not having a static accumulating tendency, such as asphalts including cutback asphalts, most crude oils, residual oils and water soluble liquids;
   (2) Where no Class I liquids are handled at the loading facililty and the tank vehicles loaded are used exclusively for Class II and Class III liquids; and
   (3) Where vehicles are loaded or unloaded through closed bottom or top connections whether the hose or pipe is conductive or nonconductive.
2. Stray Currents
   Tank car loading facilities where flammable and combustible liquids are loaded or unloaded through open domes shall be protected against stray currents by permanently bonding the pipe to at least one rail and to the rack structure, if of metal. Multiple pipes entering the rack area shall be permanently electrically bonded together. In addition, in areas where excessive stray currents are known to exist, all pipe entering the rack area shall be provided with insulating sections to electrically isolate the rack piping from the pipe lines. These precau-

tions are not necessary where Class II or Class III liquids are handled exclusively and there is no probability that tank cars will contain vapors from previous cargoes of Class I liquids. Temporary bonding is not required between the tank car and the rack or piping during either loading or unloading irrespective of the class of liquid handled.

3. Container Filling Facilities

   Class I liquids shall not be dispensed into metal containers unless the nozzle or fill pipe is in electrical contact with the container. This can be accomplished by maintaining metallic contact during filling, by a bond wire between them, or by other conductive path having an electrical resistance not greater than 106 ohms. Bonding is not required where a container is filled through a closed system, or is made of glass or other nonconducting material.

NOTE: For additional information, see ANSI/NFPA 77-1983 [33].

E. Gasoline Dispensing Stations

1. Rule 127E shall apply to areas where Class I liquids are stored, handled or dispensed. For areas where Class II or Class III liquids are stored, handled or dispensed, the electrical equipment may be installed in accordance with the provisions of applicable sections of this code.
2. All electrical equipment and wiring shall be furnished and installed in accordance with ANSI/NFPA 70-1987 [32]. All electrical equipment integral with the dispensing hose or nozzle shall be suitable for use in Division 1 locations.
3. Table 127-3 shall be used to delineate and classify areas for the purpose of installation of electrical equipment under normal circumstances. A classified area shall not extend beyond an unpierced wall, roof, or other solid partition. For a definition of the class and division designations see ANSI/NFPA 70-1987 [32], Article 500.
4. The area classifications listed in Table 127-3 are based on the premise that the installation meets the applicable requirements of this code in all respects. Should this not be the case, the authority having jurisdiction shall have the authority to determine the extent of the classified area.

F. Boilers

1. When storing, handling, or burning fuel oils which may have flash points below 100 °F (38 °C) (Class I liquids, as defined in ANSI/NFPA 30-1987 [31]), or which may be heated above their flash point, attention must be given to electrical installations in areas where flammable vapors or gases may be present in the atmosphere. Typical locations are: burner areas, fuel-handling equipment areas, fuel storage areas, pits, sumps, and low spots where fuel leakage or vapors may accumulate. ANSI/NFPA 70-1987 [32], Article 500 provides for classifying such areas and defines requirements for electrical installations in the areas so classified. The burner front piping and equipment shall be designed and constructed to eliminate hazardous concentrations of flammable gases that exist continuously, intermittently, or periodically under normal operating conditions. Providing the burners are thoroughly purged before removal for cleaning, burner front maintenance operations will not cause hazardous concentrations of flammable vapors to exist frequently. With such provisions, the burner front is not normally classified more restrictively than Class I, Division 2.
2. The operating company shall be responsible for classifying areas where fuel is stored, handled, or burned, and for revising the classification if conditions are changed. Installations shall conform to ANSI/NFPA 30-1987 [31] and ANSI/NFPA 70-1987 [32].

   NOTE: For additional guidance see API RP500A [35].

G. Gaseous Hydrogen Systems for Supply Equipment

1. Outdoor storage areas shall not be located beneath electric power lines.
2. Safety considerations at specific storage areas.

   Electrical equipment shall be suitable for Class I, Division 2 locations:

   a. Within 15 ft (4.6 m) of outdoor storage spaces
   b. Within adequately ventilated separate buildings or special rooms for storing hydrogen
   c. Within 25 ft (7.6 m) of a hydrogen storage space in an adequately ventilated building used for other purposes

3. Space around elements of the generator hydrogen seal oil system shall not be considered classified for electrical installation except where external venting is not provided in the bearing drain system.
4. Spaces around the hydrogen piping system beyond the point where the hydrogen storage system connects to distribution piping shall not be considered classified for electrical installations, outside the boundaries established in 127G2a and 127G2c.

H. Liquid Hydrogen Systems

1. Electrical wiring and equipment located within 3 ft (900 mm) of a point where connections are regularly made and disconnected shall be in accordance with ANSI/NFPA 70-1987 [32], Article 501, Class I, Group B, Division 1 locations.
2. Except as provided in Paragraph 1, electrical wiring and equipment located within 25 ft (7.6 m) of a point where connections are regularly made and disconnected or within 25 ft (7.6 m) of a liquid hydrogen storage container, shall be in accordance with ANSI/NFPA 70-1987 [32], Article 501, Class I, Group B, Division 2 locations. When equipment approved for Class I, Group B atmospheres is not commercially available, the equipment may be (1) purged or ventilated in accordance with NFPA 496-1982 [42] or (2) intrinsically safe, or (3) approved for Class I, Group C atmospheres. This requirement does not apply to electrical equipment which is installed on mobile supply trucks or tank cars from which the storage container is filled.

I. Sulfur

1. Electrical wiring and equipment located in areas where sulfur dust is in suspension in explosive or ignitable mixtures during normal operations shall be suitable for Class II, Division 1, Group G.

J. Oxygen

1. Bulk oxygen installations are not defined as classified locations.

K. Liquefied Petroleum Gas (LPG)

1. LPG is heavier than air.
2. Since LPG is contained in a closed system of piping and equipment, the system need not be electrically conductive or electrically bonded for protection against static electricity.
3. Fixed electrical equipment and wiring installed within classified areas specified in Table 127-4 shall meet the requirements of ANSI/NFPA 70-1987 [32], Article 500.

L. Natural Gas (Methane)

1. Natural Gas is lighter than air.
2. Since Natural Gas is contained in a closed system of piping and equipment, the system need not be electrically conductive or electrically bonded for protection against static electricity.
3. Fixed electrical equipment and wiring installed within classified areas specified in Table 127-5 shall meet the requirements of ANSI/NFPA 70-1987 [32], Article 500.

Table 127-1 Electrical Equipment Classified Areas — Flammable Liquid Storage Areas

Location	NEC Class I Division	Extent of Classified Area
Indoor equipment installed where flammable vapor-air mixtures may exist under normal operations.	1	Area within 5 ft (1.50 m) of any edge of such equipment, extending in all directions.
	2	Area between 5 ft (1.50 m) and 8 ft (2.45 m) of any edge of such equipment, extending in all directions. Also, area up to 3 ft (900 mm) above floor or grade level within 5 ft (1.50 m) to 25 ft (7.6 m) horizontally from any edge of such equipment.①
Outdoor equipment installed where flammable vapor-air mixtures may exist under normal operations.	1	Area within 3 ft (900 mm) of any edge of such equipment extending in all directions.
	2	Area between 3 ft (900 mm) and 8 ft (2.45 m) of any edge of such equipment extending in all directions. Also, area up to 3 ft (900 mm) above floor or grade level within 3 ft (900 mm) to 10 ft (3.0 m) horizontally from any edge of such equipment.
Tank — above ground		
Shell, ends or roof and dike area	2	Within 10 ft (3.0 m) from shell, ends or roof of tank. Area inside dikes to level of top of dike.
Vent	1	Within 5 ft (1.50 m) of open end of vent, extending in all directions
	2	Area between 5 ft (1.50 m) and 10 ft (3.0 m) from open end of vent, extending in all directions.
Floating roof	1	Area above the roof and within the shell.

① The release of Class I liquids may generate vapors to the extent that the entire building, and possibly a zone surrounding it, should be considered a Class I, Division 2 location.

(continued on next page)

Table 127-1 *(Continued)*

Location	NEC Class I Division	Extent of Classified Area
Tank — underground		
Fill opening	1	Any pit, box or space below grade level, any part of which is within the Division 1 or 2 classified area.
	2	Up to 18 in (450 mm) above grade level within a horizontal radius of 10 ft (3.0 m) from a loose fill connection and within a horizontal radius of 5 ft (1.50 m) from a tight fill connection.
Vent — discharging upward	1	Within 3 ft (900 mm) of open end of vent, extending in all directions.
	2	Area between 3 ft (900 mm) and 5 ft (1.50 m) of open end of vent, extending in all directions.
Drum and container filling; outdoors, or indoors with adequate ventilation	1	Within 3 ft (900 mm) of vent and fill opening, extending in all directions.
	2	Area between 3 ft (900 mm) and 5 ft (1.50 m) from vent or fill opening, extending in all directions. Also up to 18 in (450 mm) above floor or grade level within a horizontal radius of 10 ft (3.0 m) from vent or fill opening.

(continued on next page)

Table 127-1 *(Continued)*

Location	NEC Class I Division	Extent of Classified Area
Pumps, bleeders, withdrawal fitting, meters, and similar devices		
Indoors	2	Within 5 ft (1.50 m) of any edge of such devices, extending in all directions. Also, up to 3 ft (900 mm) above floor or grade level within 25 ft (7.6 m) horizontally from any edge of such devices.
Outdoors	2	Within 3 ft (900 mm) of any edge of such devices, extending in all directions. Also up to 18 in (450 mm) above grade level within 10 ft (3.0 m) horizontally from any edge of such devices.
Pits		
Without mechanical ventilation	2	Entire area within pit if any part is within a Division 1 or 2 classified area.
With mechanical ventilation	2	Entire area within pit if any part is within a Division 1 or 2 classified area.
Containing valves, fittings, or piping, and not within a Division 1 or 2 classified area	2	Entire Pit
Drainage ditches, separators, impounding basins	2	Area up to 18 in (450 mm) above ditch, separator, or basin. Also up to 18 in (450 mm) above grade within 15 ft (4.6 m) horizontally from any edge.

Table 127-2 Electrical Equipment Classified Areas — Bulk Plants

Location	NEC Class I, Group D Division	Extent of Classified Area
Bottom loading with vapor recovery or any bottom unloading	2	Within 3 ft (900 mm) of point of connections, extending in all directions. Also up to 18 in (450 mm) above grade within a horizontal radius of 10 ft (3.0 m) from point of connection.

Table 127-3 Electrical Equipment Classified Areas — Gasoline Dispensing Stations

Location	NEC Class I Division	Extent of Classified Area
Gasoline dispensing units (except overhead type dispensers)	1	The area up to 4 ft (1.20 m) vertically above the base within the enclosure or up to a solid partition less than 4 ft (1.20 m) above the base, located above the nozzle insertion level and above the level of any gasketed joint, hose, or stuffing box.
	2	Within 18 in (450 mm) horizontally in all directions from the Division 1 area within the enclosure.
Outdoor	2	Up to 18 in (450 mm) above grade level within 20 ft (6.1 m) horizontally of any edge of enclosure.
Indoor With mechanical ventilation	2	Up to 18 in (450 mm) above grade or floor level within 20 ft (6.1 m) horizontally of any edge of enclosure.

(continued on next page)

Table 127-3 *(Continued)*

Location	NEC Class I Division	Extent of Classified Area
With gravity ventilation	2	Up to 18 in (450 mm) above grade or floor level within 25 ft (7.6 m) horizontally of any edge of enclosure.
Gasoline dispensing units Overhead type	1	Within the dispenser enclosure and 18 in (450 mm) in all directions from the enclosure where not suitably cut off by ceiling or wall. All electrical equipment integral with the dispensing hose or nozzle.
	2	An area extending 2 ft (600 mm) horizontally in all directions beyond the Division 1 area and extending to grade below the classified area.
	2	Up to 18 in (450 mm) above grade level with 20 ft (6.1 m) horizontally measured from a point vertically below the edge of any dispenser.
Gasoline dispensing station lubrication or service room		
With dispensing	1	Any pit within any unventilated area.
	2	Any pit with ventilation.
	2	Area up to 18 in (450 mm) above floor or grade level and 3 ft (900 mm) horizontally from a lubrication pit.
Dispenser for Class I liquids	2	Within 3 ft (900 mm) of any fill or dispensing point, extending in all directions.

(continued on next page)

Table 127-3 *(Continued)*

Location	NEC Class I Division	Extent of Classified Area
Without dispensing	2	Entire area within any pit used for lubrication or similar services where Class I liquids may be released.
	2	Area up to 18 in (450 mm) above any such pit, and extending a distance of 3 ft (900 mm) horizontally from any edge of the pit.
Storage and rest rooms	Non-classified	If there is any opening to these rooms within the extent of a Division 1 area, the entire room shall be classified as Division 1.

Location	NEC Class I, Group D Division	Extent of Classified Area
Vapor processing pits	1	Any pit, box, or space below grade level, any part of which is within a Division 1 or 2 location or which houses any equipment used to transport or process vapors.
Equipment	2	Within protective enclosures. The space within 18 in (450 mm) in all directions of equipment containing flammable vapor or liquid extending to grade level. Up to 18 in (450 mm) above grade level within 10 ft (3.0 m) horizontally of the vapor processing equipment.

Table 127-4 Electrical Equipment Classified Areas — LPG Storage

Location	NEC Class I Group D	Extent of Classified Area
Storage containers other than DOT cylinders	2	Within 15 ft (4.6 m) in all directions from connections, except connections otherwise covered in Table 127-4.
Tank vehicle and tank car loading and unloading	1	Within 5 ft (1.50 m) in all directions from connections regularly made or disconnected for product transfer.
	2	Beyond 5 ft (1.50 m) but within 15 ft (4.6 m) in all directions from a point where connections are regularly made or disconnected and within the cylindrical volume between the horizontal equator of the sphere and grade.
Gage vent openings other than those on DOT cylinders	1	Within 5 ft (1.50 m) in all directions from point of discharge.
	2	Beyond 5 ft (1.50 m) but within 15 ft (4.6 m) in all directions from point of discharge.
Relief valve discharge other than those on DOT cylinders	1	Within direct path of discharge. *Note:* Fixed electrical equipment should preferably not be installed.
	1	Within 5 ft (1.50 m) in all directions from point of discharge.
	2	Beyond 5 ft (1.50 m) but within 15 ft (4.6 m) in all directions from point of discharge except within the path of discharge.

(continued on next page)

Table 127-4 *(Continued)*

Location	NEC Class I Group D	Extent of Classified Area
Pits or trenches containing or located beneath LP-Gas valves, regulators, and similar equipment		
Without mechanical ventilation	1	Entire pit or trench.
	2	Entire room and any adjacent room not separated by a gastight partition.
	2	Within 15 ft (4.6 m) in all directions from pit or trench when located outdoors.
With adequate mechanical ventilation	2	Entire pit or trench.
	2	Entire room and any adjacent room not separated by a gastight partition.
	2	Within 15 ft (4.6 m) in all directions from pit or trench when located outdoors.
Special buildings or rooms for storage of portable containers	2	Entire room.
Pipelines and connections containing operational bleeds, drips, vents, or drains	1	Within 5 ft (1.50 m) in all directions from point of discharge.
Container filling:		
Indoors without ventilation	1	Entire room.
Indoors with adequate ventilation	1	Within 5 ft (1.50 m) in all directtions and connections regularly made or disconnected for product transfer.
	2	Beyond 5 ft (1.50 m) and entire room.

(continued on next page)

Table 127-4 *(Continued)*

Location	NEC Class I Group D	Extent of Classified Area
Container filling		
Outdoors in open air	1	Within 5 ft (1.50 m) in all directions and connections regularly made or disconnected for product transfer.
	2	Beyond 5 ft (1.50 m) but within 15 ft (4.6 m) in all directions from a point where connections are regularly made or disconnected and within the cylindrical volume between the horizontal equator of the sphere and grade.

Table 127-5 Electrical Equipment Classified Areas — Natural Gas (Methane) Areas

Location	NEC Class I Group D	Extent of Classified Area
Nonfired areas containing gas pipeline connections, valves, or gages		
Indoors with adequate ventilation	2	Entire room and any adjacent room not separated by a gastight partition and 15 ft (4.6 m) beyond any wall or roof ventilation discharge vent or louver.
Outdoors in open air at or above grade	2	Within 15 ft (4.6 m) in all directions of connections, valves, or gages.
Pits, trenches, or sumps located in or adjacent to Division 1 or 2 areas	1	Entire pit, trench, or sump.

128. Identification

Electrical equipment and devices shall be identified for safe use and operation. The identification shall be as nearly uniform as practical throughout any one station. Identification marks shall not be placed on removable covers or doors where the interchanging of those covers or doors is possible.

129. Mobile Hydrogen Equipment

Mobile hydrogen supply units being used to replenish a hydrogen system shall be bonded both to the grounding system and to the grounded parts of the hydrogen system.

Section 13. Rotating Equipment

Rotating equipment includes generators, motors, motor generators, and rotary converters.

130. Speed Control and Stopping Devices

A. Automatic Overspeed Trip Device for Prime Movers

When harmful overspeed can occur, prime movers driving generating equipment shall be provided with automatic overspeed trip devices in addition to their governors.

B. Manual Stopping Devices

An operator initiated stopping device shall be provided for any machine that drives an electric power generator or rotary uninterruptible power supply (motor-generator). The operator initiated stopping device shall be accessible to the operator during normal operation. Manual controls to be used in emergency for machinery and electrical equipment shall be located so as to provide protection to the operator in the event of such emergency.

C. Speed Limit for Motors

Machines of the following types shall be provided with speed-limiting devices unless their inherent characteristics or the load and the mechanical connection thereto are such as to safely limit the speed.

1. Separately excited direct-current motors.
2. Series motors.

D. Number 130D not used in this edition.

E. Adjustable-Speed Motors

Adjustable-speed motors, controlled by means of field regulation, shall, in addition to the provisions of Rule 130C, be so equipped and connected that the field cannot be weakened sufficiently to permit dangerous speed.

F. Protection of Control Circuits

Where speed-limiting or stopping devices and systems are electrically operated, the control circuits by which such devices are actuated shall be protected from mechanical damage. Such devices and systems should be of the automatic tripping type.

131. Motor Control

If the starting is automatic, as for example, by a float switch, or if the starting device or control switch is not in sight, or more than 50 ft (15 m) distant from the motor and all parts of the machinery operated, the power or control circuit shall be such that it can positively be kept open as by use of lockout/tagout procedures.

All motors so employed or arranged that an unexpected starting of the motor might create an exposure of personnel to injury shall have the motor control designed to prevent automatic reenergization of the motor after a power supply interruption of a duration sufficient for moving equipment to become stationary. The motor control shall be such that an operator must take some action to restart the motor, or automatic restarting shall be preceded by warning signals and a time delay sufficient for personnel action to prevent injury. This requirement does not apply to those motors with an emergency use and where the opening of the circuit may cause less safe conditions.

132. Number 132 not used in this edition.

133. Short Circuit Protection

Means shall be provided to automatically disconnect an electric motor from the supply source in the event of high magnitude short-circuit currents within the motor.

Section 14. Storage Batteries

140. General

The provisions of this section are intended to apply to all stationary installations of storage batteries. For operating precautions, see Part 4 of this code.

Space shall be provided around batteries for safe inspection, maintenance, testing, and cell replacement and space left above the cells to allow for operation of lifting equipment when required, addition of water, and taking measurements.

141. Location

Storage batteries shall be located within a protective enclosure or area accessible only to qualified persons. A protective enclosure can be a battery room, control building, or a case, cage, or fence which will protect the contained equipment and minimize the possibility of inadvertent contact with energized parts.

142. Ventilation

The battery area shall be ventilated, either by a natural or powered ventilation system to prevent accumulation of hydrogen. The ventilation system shall limit hydrogen accumulation to less than an explosive mixture. Failure of continuously operated or automatically controlled powered ventilation system shall be annunciated.

143. Racks

Racks refer to frames designed to support cells or trays. Racks shall be firmly anchored preferably to the floor. Anchoring to both walls and floors is not recommended. Racks made of metal shall be grounded.

144. Floors in Battery Areas

Floors of battery areas should be an acid-resistive material, or be painted with acid-resistive paint, or otherwise protected. Provision should be made to contain spilled electrolyte.

145. Illumination for Battery Areas

Lighting fixtures shall be protected from physical damage by guards or isolation. Receptacles and lighting switches should be located outside of battery areas.

146. Service Facilities

A. Proper eye protection and clothing shall be provided in the battery area during battery maintenance and installation and shall consist of:

1. Goggles or face shield
2. Acid resistant gloves
3. Protective aprons and overshoes
4. Portable or stationary water facilities or neutralizing agent for rinsing eyes and skin

B. Warning signs inside and outside of a battery room or in the vicinity of a battery area, prohibiting smoking, sparks, or flame shall be provided.

147. Number 147 not used in this edition.

Section 15. Transformers and Regulators

150. Current-Transformer Secondary Circuits Protection When Exceeding 600 Volts

Secondary circuits, when in a primary voltage area exceeding 600 V should, except for short lead lengths at the terminals of the transformer, have the secondary wiring adequately protected by means of grounded conduit or by a grounded metallic covering. Current transformers shall have provision for shorting the secondary winding.

151. Grounding Secondary Circuits of Instrument Transformers

The secondary circuits of instrument transformers shall be effectively grounded where functional requirements permit.

152. Location and Arrangement of Power Transformers and Regulators

A. Outdoor Installations

1. A transformer or regulator shall be installed so that all energized parts are enclosed or guarded so as to minimize the possibility of inadvertent contact, or the ener-

gized parts shall be isolated in accordance with Rule 124. The case shall be grounded in accordance with Rule 123.

2. Oil-filled transformers shall be protected by one or more of the following methods to minimize fire hazards. The method to be applied shall be according to the degree of fire hazard and the amount of oil contained in the transformer. Recognized methods are space separation, fire-resistant barriers, automatic extinguishing systems, absorption beds and enclosures.

   The amount of oil contained should be considered in the selection of space separation, fire resistant barriers, automatic extinguishing systems, absorption beds, and enclosures which confine the oil of a ruptured transformer tank, all of which are recognized as safeguards.

B. Indoor Installations

1. Transformers and regulators 75 kVA and above containing an appreciable amount of flammable liquid and located indoors shall be installed in ventilated rooms or vaults separated from the balance of the building by fire walls. Doorways to the interior of the building shall be equipped with fire doors and shall have means of containing the oil.
2. Transformers or regulators of the dry type or containing a non-flammable liquid or gas may be installed in a building without a fireproof enclosure. When installed in a building which is used for other than station purposes the case or the enclosure shall be designed so that all energized parts are enclosed in the case grounded in accordance with Rule 123. As an alternate, the entire unit may be enclosed so as to minimize the possibility of inadvertent contact by persons with any part of the case or wiring. When installed, the pressure relief vent of a unit containing a non-biodegradable liquid shall be furnished with a means for absorbing toxic gases.

Section 16. Conductors

160. Electrical Protection

Conductors shall be suitable for the location, use, and voltage.

A. Overcurrent Protection Required

Conductors and insulation shall be protected against excessive heating by the design of the system and by overcurrent, alarm, indication, or trip devices.

B. Grounded Conductors

Conductors normally grounded for the protection of persons shall be arranged without overcurrent protection or other means which could interrupt their continuity to ground.

C. Insulated Power Cables

Insulated power cable circuits shall be provided with short-circuit protection that will isolate the short circuit from the supply.

161. Mechanical Protection

All conductors shall be adequately supported to withstand forces caused by the maximum short circuit current to which they may be subjected.

Where exposed to mechanical damage, casing, armor, or other means shall be employed to prevent damage or disturbance to conductors, their insulation, or their supports.

162. Isolation

All non-shielded insulated conductors of more than 2500 volts to ground and bare conductors of more than 150 V to ground, shall be isolated by elevation or guarded in accordance with Rule 124.

Non-shielded, insulated, and jacketed conductors may be installed in accordance with Rule 124C6.

163. Conductor Terminations

A. Insulation

Ends and joints of insulated conductors, unless otherwise adequately guarded, shall have insulating covering equivalent to that of other portions of the conductor.

B. Metal-Sheathed or Shielded Cable

Insulation of the conductors where leaving the metal sheath or shield, shall be protected from mechanical damage, moisture and excessive electrical stress.

Section 17. Circuit Breakers, Reclosers, Switches, and Fuses

170. Arrangement

Circuit breakers, reclosers, switches and fuses shall be so installed as to be accessible only to persons qualified for operation and maintenance. Walls, barriers, latched doors, location, isolation or other means shall be provided to protect persons from energized parts or arcing. Conspicuous marking shall be provided at the device and at any remote operating points to identify the equipment controlled. When the contact parts of a switching device are not normally visible, the device shall be equipped with an indicator to show all normal operating positions.

171. Application

Circuit breakers, reclosers, switches, and fuses should be utilized with due regard to their assigned ratings of voltage and continuous and momentary currents. Circuit breakers, reclosers and fuses which perform a fault current interrupting function shall be capable of safely interrupting the maximum short circuit current available from the system at the point of application. The interrupting capacity should be reviewed prior to each significant system change.

172. Circuit Breakers, Reclosers and Switches Containing Oil

Circuit interrupting devices containing flammable liquids shall be adequately segregated from other equipment and buildings to limit damage in the event of an explosion or fire. Segregation may be provided by spacing, by fire-resistant barrier walls, or by metal cubicles. Gas relief vents should be equipped with oil separating devices or piped to a safe location. Means shall be provided to control oil which could be discharged from vents or by tank rupture. This may be accomplished by absorption beds, pits, drains, or by any combination of these. Buildings or rooms housing this equipment shall be of fire resistant construction.

173. Switches and Disconnecting Devices

A. Capacity

Switches shall be of suitable voltage and ampere rating for the circuit in which they are installed. Switches used to break load current shall be marked with the current which they are rated to interrupt.

B. Provisions for Disconnecting

Switches and disconnectors shall be so arranged that they can be locked in the open and closed positions, or plainly tagged where it is not possible to install locks. For devices that are operated remotely and automatically, the control circuit shall be provided with a positive disconnecting means near the apparatus to prevent accidental operation of the mechanism.

C. Visible Break Switch

A visible break switch or disconnector shall be inserted in each ungrounded conductor between electric supply equipment or lines and sources of energy of more than 600 V, if the equipment or lines may have to be worked on without protective grounding while the sources may be energized.

Where metal clad switchgear equipment is used, the withdrawn position of the circuit breaker, where clearly indicated, constitutes a visible break for this purpose.

174. Disconnection of Fuses

Fuses in circuits of more than 150 V to ground or more than 60 A shall be classified as disconnecting fuses or be arranged so that before handling:

A. The fuses can be disconnected from all sources of electric energy, or

B. The fuses can be conveniently removed by means of insulating handles.

Fuses can be used to disconnect from the source when they are so rated.

Section 18. Switchgear and Metal Enclosed Bus

180. Switchgear Assemblies

A. General Requirements for All Switchgear

1. To minimize movement, all switchgear shall be secured in a manner consistent with its conditions of service and applicable manufacturer's instructions.

2. Cable routed to switchgear shall be supported to minimize forces applied to conductor terminals.
3. Piping containing liquids, or corrosive or hazardous gases, shall not be routed in the vicinity of switchgear unless suitable barriers are installed to protect the switchgear from damage in the event of a pipe failure.
4. Switchgear shall not be located where foreign flammable or corrosive gases or liquids routinely and normally are discharged. Companion equipment such as transformers and switchgear are not considered foreign.
5. Switchgear should not be installed in a location which is still specifically under active construction, especially where welding and burning are required directly over head. Special precautions should be observed to minimize impingement of slag, metal filings, moisture, dust, or hot particles.

   EXCEPTION: Switchgear may be installed in a general construction area provided suitable temporary protection is provided to minimize the risks associated with general construction activities.
6. Precautions shall be taken to protect energized switchgear from damage when maintenance is performed in the area.
7. Switchgear enclosure surfaces shall not be used as physical support for any item unless specifically designed for that purpose.
8. Enclosure interiors shall not be used as storage areas unless specifically designed for that purpose.
9. Metal instrument cases shall be grounded, enclosed in covers which are metal and grounded, or of insulating material.

B. Metal Enclosed Power Switchgear

1. Switchgear shall not be located within 25 ft (7.6 m) horizontally indoors or 10 ft (3.0 m) outdoors of storage containers, vessels, utilization equipment or devices containing flammable liquids or gases.

   EXCEPTION: If an intervening barrier, designed to mitigate the potential effects of flammable liquids or gases, is installed, the distances listed above do not apply.

   The restrictions are not intended to apply to the power transformer(s) supplying the switchgear.

2. Enclosed switchgear rooms shall have at least two means of egress, one at each extreme of the area, not necessarily in opposite walls. Doors shall swing out and be equipped with panic bars, pressure plates or other devices that are normally latched but open under simple pressure.

   EXCEPTION: One door may be used when required by physical limitations if means are provided for unhampered exit during emergencies.
3. Space shall be maintained in front of switchgear to allow breakers to be removed and turned without obstruction.
4. Space shall be maintained in the rear of the switchgear to allow for door opening to at least 90 degrees open, or a minimum of 3 feet and no inches (900 mm) without obstruction when removable panels are used.
5. Permanently mounted devices, panelboards, etc located on the walls shall not encroach on the space requirements in 180B4.
6. Where columns extend into the room beyond the wall surface, the face of the column shall not encroach on the space requirements in 180B4.
7. Low-voltage cables or conductors, except those to be connected to equipment within the compartment, shall not be routed through the medium or high-voltage divisions of switchgear unless installed in rigid metal conduit or isolated by rigid metal barriers.
8. Low voltage conductors routed from medium or high voltage sections of switchgear shall terminate in a low voltage section before being routed external to the switchgear.
9. Conductors entering switchgear shall be insulated for the higher operating voltage in that compartment or be separated from insulated conductors of other voltage ratings.
10. Switchgear enclosures shall be suitable for the environment in which they are installed.
11. A warning sign shall be placed in each cubicle containing more than one high voltage source.

12. The location of control devices shall be readily accessible to personnel. Instruments, relays and other devices requiring reading or adjustments should be so placed that work can readily be performed from the working space.

C. Dead Front Power Switchboards

Dead front power switchboards with uninsulated rear connections shall be installed in rooms or spaces that are capable of being locked, with access limited to qualified personnel.

D. Motor Control Centers

1. Motor control centers shall not be connected to systems having higher short circuit capability than the bus bracing can withstand. Where current limiting fuses are employed on the source side of the bus, the bus bracing and breaker interrupting rating are determined by the peak let-through characteristic of the current limiting fuse.
2. A warning sign shall be placed in each cubicle containing more than one voltage source.

E. Control Switchboards

1. Cabinets containing solid-state logic devices, electron tubes or relay logic devices such as boiler analog, burner safety, annunciators, computers, inverters, precipitator logic, soot blower control, load control, telemetering, totalizing microwave radio, etc are covered under these rules.
2. Where carpeting is installed in rooms containing control switchboards, it shall be antistatic type and shall minimize the release of noxious, corrosive, caustic, or toxic gas under any condition.
3. Layout of the installation shall provide adequate clearance in front of, or rear of panels if applicable, to allow meters to be read without use of stools or auxiliary devices.
4. Where personnel access to control panels such as benchboards is required, cables shall be routed through openings separate from the personnel opening. Removable, sliding, or hinged panels are to be installed to close the personnel opening when not in use.

181. Metal Enclosed Bus

A. General Requirements for All Types of Bus

1. Busways shall be installed only in accessible areas.
2. Busways unless specifically approved for the purpose, shall not be installed: where subject to severe physical damage or corrosive vapors; in hoistways; in any classified hazardous location; outdoors or in damp locations.
3. Dead ends of busway shall be closed.
4. Busways should be marked with the voltage and current rating for which they are designed, in such manner as to be visible after installation.

B. Isolated Phase Bus

1. The minimum clearance between an isolated-phase bus and any magnetic material shall be the distance recommended by the manufacturer to avoid overheating of the magnetic material.
2. Non-magnetic conduit should be used to protect the conductors for bus alarm devices, thermocouples, space heaters, etc if routed within the manufacturer's recommended minimum distance to magnetic material and parallel to isolated-phase bus enclosures.
3. When enclosure drains are provided for isolated-phase bus, necessary piping shall be provided to divert water away from electrical equipment.
4. Wall plates for isolated-phase bus shall be non-magnetic, such as aluminum or stainless steel.
5. Grounding conductors for isolated-phase bus accessories should not be routed through ferrous conduit.

Section 19. Surge Arresters

190. General Requirements

If arresters are required, they shall be located as close as practical to the equipment they protect.

191. Indoor Locations

Arresters, if installed inside of buildings shall be enclosed or shall be located well away from passageways and combustible parts.

192. Grounding Conductors

Grounding conductors shall be run as directly as possible between the arresters and ground and be of low impedance and ample current-carrying capacity (see Section 9 for methods of protective grounding).

193. Installation

Arresters shall be installed in such a manner and location that neither the expulsion of gases nor the arrester disconnector is directed upon live parts in the vicinity.

PART 2. Safety Rules for the Installation and Maintenance of Overhead Electric Supply and Communication Lines

Section 20. Purpose, Scope, and Application of Rules

200. Purpose

The purpose of Part 2 of this code is the practical safeguarding of persons during the installation, operation, or maintenance of overhead supply and communication lines and their associated equipment.

201. Scope

Part 2 of this code covers supply and communication conductors and equipment in overhead lines. It covers the associated structural arrangements of such systems and the extension of such systems into buildings. The rules include requirements for spacing, clearances, and strength of construction. They do not cover installations in electric supply stations.

202. Application of Rules

The general requirements for application of these rules are contained in Rule 013. However, when a structure is replaced, the arrangement of equipment shall conform to the current edition of Rule 238C.

Section 21. General Requirements

210. Referenced Sections

The Introduction (Section 1), Definitions (Section 2), References (Section 3) and Grounding Methods (Section 9) shall apply to the requirements of Part 2.

211. Number 211 not used in this edition.

212. Induced Voltages

Rules covering supply line influence and communication line susceptiveness have not been detailed in this code. Cooperative procedures are recommended in the control of voltages induced from proximate facilities. Therefore, reasonable advance notice should be given to owners or operators of other proximate facilities which may be adversely affected by new construction or changes in existing facilities.

213. Accessibility

All parts which must be examined or adjusted during operation shall be arranged so as to be accessible to authorized persons by the provision of adequate climbing spaces, working spaces, working facilities, and clearances between conductors.

214. Inspection and Tests of Lines and Equipment

A. When In Service

1. Initial Compliance With Rules
   Lines and equipment shall comply with these safety rules when placed in service.
2. Inspection
   Lines and equipment shall be inspected at such intervals as experience has shown to be necessary.
3. Tests
   When considered necessary, lines and equipment shall be subjected to practical tests to determine required maintenance.
4. Record of Defects
   Any defects affecting compliance with this code revealed by inspection or tests, if not promptly corrected, shall be recorded; such records shall be maintained until the defects are corrected.
5. Remedying Defects
   Lines and equipment with recorded defects which could reasonably be expected to endanger life or property shall be promptly repaired, disconnected, or isolated.

B. When Out of Service

1. Lines Infrequently Used

Lines and equipment infrequently used shall be inspected or tested as necessary before being placed into service.

2. Lines Temporarily Out of Service

Lines and equipment temporarily out of service shall be maintained in a safe condition.

3. Lines Permanently Abandoned

Lines and equipment permanently abandoned shall be removed or maintained in a safe condition.

215. Grounding of Circuits, Supporting Structures, and Equipment

A. Methods

Grounding required by these rules shall be in accordance with the applicable methods given in Section 9.

B. Circuits

1. Common Neutral

A conductor used as a common neutral for primary and secondary circuits shall be effectively grounded as specified in Section 9.

2. Other Neutrals

Primary line, secondary line, and service neutral conductors shall be grounded as specified in Section 9.

EXCEPTION: Circuits designed for ground fault detection and impedance current limiting devices.

3. Other Conductors

Line or service conductors, other than neutral conductors, which are intentionally grounded, shall be grounded as specified in Section 9.

4. Surge Arresters

Where the operation of surge arresters is dependent upon grounding, they shall be grounded in accordance with the methods outlined in Section 9.

5. Use of Earth as Part of Circuit

Supply circuits shall not be designed to use the earth normally as the sole conductor for any part of the circuit.

C. Noncurrent-Carrying Parts

1. General

Metal or metal reinforced supporting structures, including lamp posts; metal conduits and raceways; cable sheaths; messengers; metal frames, cases and hangers of equipment; and metal switch handles and operating rods shall be effectively grounded.

EXCEPTION 1: This rule does not apply to frames, cases, and hangers of equipment and switch handles and operating rods which are 8 ft (2.45 m) or more above readily accessible surfaces or are otherwise isolated or guarded and where the practice of not grounding such items has been a uniform practice over a well defined area.

EXCEPTION 2: This rule does not apply to isolated or guarded equipment cases in certain specialized applications, such as series capacitors where it is necessary that equipment cases be either ungrounded or connected to the circuit. Such equipment cases shall be considered as energized and shall be suitably identified.

EXCEPTION 3: This rule does not apply to equipment cases, frames, equipment hangers, conduits, raceways, and cable sheaths enclosing only communications conductors, provided they are not exposed to probable contact with open supply conductors of over 300 volts.

2. Guys

Guys shall be effectively grounded if attached to a supporting structure carrying any supply conductor of more than 300 volts or if exposed to such conductors.

EXCEPTION 1: This rule does not apply to guys containing an insulator or insulators installed in accordance with and meeting the requirements of Rule 279A.

EXCEPTION 2: This rule does not apply to guys attached to supporting structures if all supply conductors are in cable conforming to the requirements of Rules 230C1, 230C2, or 230C3.

EXCEPTION 3: This rule does not apply if the guy is attached to a supporting structure on private right-of-way if all the supply circuits exceeding 300 volts meet the requirements of Rule 220B2.

3. Multiple Messengers on the Same Structure

Communication cable messengers exposed to power contacts, power induction, or lightning, shall be bonded together at intervals specified in Rule 92C.

216. Arrangement of Switches

A. Accessibility

Switches or their control mechanisms shall be installed so as to be accessible to authorized persons.

B. Indicating Open or Closed Position

Switch position shall be visible or clearly indicated.

C. Locking

Switch operating mechanisms which are accessible to unauthorized persons shall have provisions for locking in each operational position.

D. Uniform Position

The handles or control mechanisms for all switches throughout any system should have consistent positions when opened and uniformly different positions when closed in order to minimize operating errors. Where this practice is not followed, the switches should be marked to minimize mistakes in operation.

217. General

A. Supporting Structures

1. Protection of Structures

a. Mechanical Injury

Appropriate physical protection shall be provided for supporting structures subject to vehicular traffic abrasion which would materially affect their strength.

b. Climbing

Readily climbable supporting structures, such as closely latticed poles or towers, including those attached to bridges, carrying open supply conductors energized at more than 300 volts, which are adjacent to roads, regularly travelled pedestrian thoroughfares, or places where persons frequently gather (such as schools or public playgrounds) shall be equipped with barriers to inhibit climbing by unqualified persons or posted with appropriate warning signs.

EXCEPTION: This rule does not apply where the right-of-way is fenced.

c. Fire
Supporting structures shall be placed and maintained so as to be exposed as little as is practical to brush, grass, rubbish, or building fires.

d. Attached to Bridges
Supporting structures attached to bridges for the purpose of carrying open supply conductors exceeding 600 volts shall be posted with appropriate warning signs.

2. Steps
Steps permanently installed on supporting structures shall not be closer than 8 ft (2.45 m) from the ground or other accessible surface.
EXCEPTION: This rule does not apply where supporting structures are isolated.

3. Identification
Supporting structures, including those on bridges, on which supply or communication conductors are maintained shall be so constructed, located, marked, or numbered so as to facilitate identification by employees authorized to work thereon. Date of installation of such structures should be recorded where practical by the owner.

4. Obstructions
Signs, posters, notices, and other attachments shall not be placed on supporting stuctures without concurrence of the owner. Supporting structures should be kept free from other climbing hazards such as tacks, nails, vines, and through bolts not properly trimmed.

5. Decorative Lighting
Attachment of decorative lighting on structures shall not be made without the concurrence of the owners and occupants.

B. Unusual Conductor Supports
Where conductors are attached to structures other than those used solely or principally for their support, all rules shall be complied with as far as they apply. Such additional precautions as may be deemed necessary by the administrative authority shall be taken to avoid damage to the structures or injury to the persons using them. The supporting of conductors on trees and roofs should be avoided.

218. Tree Trimming

A. General

1. Trees which may interfere with ungrounded supply conductors should be trimmed or removed.

   NOTE: Normal tree growth, the combined movement of trees and conductors under adverse weather conditions, voltage, and sagging of conductors at elevated temperatures are among the factors to be considered in determining the extent of trimming required.

2. Where trimming or removal is not practical, the conductor should be separated from the tree with suitable materials or devices to avoid conductor damage by abrasion and grounding of the circuit through the tree.

B. At Line Crossings, Railroad Crossings, and Limited Access Highway Crossings

The crossing span and the adjoining span on each side of the crossing should be kept free from overhanging or decayed trees or limbs which otherwise might fall into the line.

Section 22. Relations Between Various Classes of Lines and Equipment

220. Relative Levels

A. Standardization of Levels

The levels at which different classes of conductors are to be located should be standarized by agreement of the utilities concerned.

B. Relative Levels: Supply and Communication Conductors

1. Preferred Levels

   Where supply and communication conductors cross each other or are located on the same structures, the supply conductors should be carried at the higher level.

   EXCEPTION: This rule does not apply to trolley feeders which may be located for convenience approximately at the level of the trolley-contact conductor.

2. Special Construction for Supply Circuits, the Voltage of Which is 600 Volts or Less and Carrying Power Not in Excess of 5 Kilowatts.

   Where all circuits are owned or operated by one party or where cooperative consideration determines that the circumstances warrant and the necessary coordinating methods are employed, single-phase alternating-current or two-wire direct-current circuits carrying a voltage of 600 volts or less between conductors, with transmitted power not in excess of 5 kilowatts, when involved in the joint use of structures with communication circuits may be installed in accordance with footnote 1 of Table 235-5, under the following conditions.

a. That such supply circuits are of covered conductor not smaller than AWG No 8 medium hard-drawn copper or its equivalent in strength, and the construction otherwise conforms with the requirements for supply circuits of the same class.

b. That the supply circuits be placed on the end and adjacent pins of the lowest through signal support arm and that a 30 inch (750 mm) climbing space be maintained from the ground up to a point at least 24 in (600 mm) above the supply circuits. The supply circuits shall be rendered conspicuous by the use of insulators of different form or color from others on the pole line or by stenciling the voltage on each side of the support arm between the pins carrying each supply circuit, or by indicating the voltage by means of metal characters.

c. That there shall be a vertical clearance of at least 2 ft (600 mm) between the support arm carrying these supply circuits and the next support arm above. The other pins on the support arm carrying the supply circuits may be occupied by communication circuits used in the operation or control of signal system or other supply system if owned, operated, and maintained by the same company operating the supply circuits.

d. That such supply circuits shall be equipped with arresters and fuses installed in the supply end of the circuit and where the signal circuit is alternating current, the protection shall be installed on the secondary side of the supply transformer. The arresters shall be designed so as to break down at approximately twice the voltage between the wires of the circuit, but the breakdown voltage of the arrester need not be less than 1 kilovolt. The fuses shall have a rating not in excess of approximately twice the maximum operating current of the circuit, but their rating need not be less than 10 amperes. The fuses likewise shall in all cases have a rating of at least 600 volts, and where the supply transformer is a step-down transformer, shall be capable of opening the circuit successfully in the event the transformer primary voltage is impressed upon them.

e. Such supply circuits in cable meeting the requirements of Rules 230C1, 230C2, or 230C3 may be installed below communication attachments, with not less than 2 ft (600 mm) vertical separation between the supply cable and the lowest communication attachment. Communication circuits other than those used in connection with the operation of the supply circuits shall not be carried in the same cable with such supply circuits.

f. Where such supply conductors are carried below communication conductors, transformers and other apparatus associated therewith shall be attached only to the sides of the support arm in the space between, and at no higher level than, such supply wires.

g. Lateral runs of such supply circuits carried in a position below the communication space shall be protected through the climbing space by wood molding or equivalent covering, or shall be carried in insulated multiple-conductor cable, and such lateral runs shall be placed on the under side of the support arm.

C. Relative Levels: Supply Lines of Different Voltage Classifications (as classified in Table 235-5)

1. At Crossings or Conflicts

Where supply conductors of different voltage classifications cross each other or structure conflict exists, the higher voltage lines should be carried at the higher level.

2. On Structures Used Only by Supply Conductors
   Where supply conductors of different voltage classifications are on the same structures, relative levels should be as follows:
   a. Where all circuits are owned by one utility, the conductors of higher voltages should be placed above those of lower voltage.
   b. Where different circuits are owned by separate utilities, the circuits of each utility may be grouped together and one group of circuits may be placed above the other group provided that the circuits in each group are located so that those of higher voltage are at the higher levels and that any of the following conditions is met:
      (1) A vertical spacing of not less than that required by Table 235-5 is maintained between the nearest line conductors of the respective utilities.
      (2) Conductors of a lower voltage classification placed at a higher level than those of a higher classification shall be placed on the opposite side of the structure.
      (3) Ownership and voltage are prominently displayed.

D. Identification of Overhead Conductors
All conductors of electric-supply and communication lines should, as far as is practical, be arranged to occupy uniform positions throughout, or shall be constructed, located, marked, numbered, or attached to distinctive insulators or crossarms, so as to facilitate identification by employees authorized to work thereon. This does not prohibit systematic transposition of conductors.

E. Identification of Equipment on Supporting Structures
All equipment of electric-supply and communication lines should be arranged to occupy uniform positions throughout or shall be constructed, located, marked, or numbered so as to facilitate identification by employees authorized to work thereon.

221. Avoidance of Conflict

Two separate lines, either of which carries supply conductors, should be so separated from each other that neither conflicts with the other. If this is not practical, the conflicting line or lines should be separated as far as possible and shall be built to the grade of construction required by Section 24 for a conflicting line, or the two lines shall be combined on the same structures.

222. Joint Use of Structures

Joint use of structures should be considered for circuits along highways, roads, streets, and alleys. The choice between joint use of structures and separate lines shall be determined through cooperative consideration of all the factors involved, including the character of circuits, the total number and weight of conductors, tree conditions, number and location of branches and service drops, possible structure conflicts, availability of right-of-way, etc. Where such joint use is mutually agreed upon, it shall be subject to the appropriate grade of construction specified in Section 24.

223. Communications Protective Requirements

A. Where Required

Where communications apparatus is handled by other than qualified persons, it shall be protected by one or more of the means listed in Rule 223B if such apparatus is permanently connected to lines subject to any of the following:

1. Lightning
2. Possible contact with supply conductors whose voltage to ground exceeds 300 volts
3. Transient rise in ground potential exceeding 300 volts
4. Steady state induced voltage of a hazardous level

   NOTE: When communication cables will be in the vicinity of supply stations where large ground currents may flow, the effect of these currents on communications circuits should be evaluated.

B. Means of Protection

Where communication apparatus is required to be protected under Rule 223A, protective means adequate to withstand the voltage expected to be impressed shall be provided by insulation, protected where necessary by surge arresters used in conjunction with fusible elements. Severe conditions may require the use of additional devices such as auxiliary arresters, drainage coils, neutralizing transformers, or isolating devices.

224. Circuits of One Class Used Exclusively in the Operation of Circuits of Another Class

A. Overhead Communication Circuits Used Exclusively in the Operation of Supply Circuits

1. Communication circuits used exclusively in the operation of supply lines may be run either as ordinary communication circuits or as supply circuits under the conditions specified in Rules 224A3 and 224A4, respectively. After the selection of the type of communication circuit construction and protection for a section, such construction and protection shall be consistently adhered to throughout the extent of such section of the communication system.
2. Communication circuits used in operation of supply lines shall be isolated or guarded at all points so as to be inaccessible to the public.
3. Communication circuits used in operation of supply lines may be run as ordinary communication conductors under the following conditions:
   a. Where such circuits are below supply conductors in the operation of which they are used (including high-voltage trolley feeders) at crossings, conflicts, or on commonly used poles, provided:
      (1) Such communication circuits occupy a position below all other supply conductors or equipment at crossings, conflicts, or on commonly used poles
      (2) Such communication circuits and their connected equipment are adequately guarded and are accessible only to authorized persons
   b. Where such circuits are below supply conductors in the operation of which they are used and are above other supply or communication conductors at wire crossings, conflicts, or on the same poles, provided

the communication circuits are protected by fuseless surge arresters, drainage coils, or other suitable devices to prevent the communication circuit voltage from normally exceeding 400 volts to ground.

NOTE: The grades of construction for communication conductors with inverted levels apply.

4. Communication circuits used in the operation of supply lines shall comply with all requirements for the supply lines with which they are used, where they do not comply with the provisions of 3a or 3b above.

   EXCEPTION 1: If the voltage of the supply conductors concerned exceeds 8.7 kilovolts, the communication conductors need only meet the requirements for supply conductors of 5 to 8.7 kilovolts.

   EXCEPTION 2: Where the supply conductors are required to meet grade C, the size of the communication conductors may be the same as for grade D (see Rule 262I2) for spans up to 150 ft (45 m).

B. Supply Circuits Used Exclusively in the Operation of Communication Circuits

Circuits used for supplying power solely to apparatus forming part of a communications system shall be installed as follows:

1. Open wire circuits shall have the grades of construction, clearances, insulation, etc, prescribed elsewhere in these rules for supply or communication circuits of the voltage concerned.
2. Special circuits operating at voltages in excess of 400 volts to ground and used for supplying power solely to communications equipment may be included in communications cables under the following conditions:
   a. Such cables shall have a conductive sheath or shield which is effectively grounded and each such circuit shall be carried on conductors which are individually enclosed with an effectively grounded shield.
   b. All circuits in such cables shall be owned or operated by one party and shall be maintained only by qualified personnel.
   c. Supply circuits included in such cables shall be terminated at points accessible only to qualified personnel.

d. Communications circuits brought out of such cables, if they do not terminate in a repeater station or terminal office, shall be protected or arranged so that in the event of failure within the cable, the voltage on the communication circuit will not exceed 400 volts to ground.

e. Terminal apparatus for the power supply shall be so arranged that the live parts are inaccessible when such supply circuits are energized.

EXCEPTION: The requirements of Rule 224 do not apply to the supply circuits of 600 volts or less where the transmitted power does not exceed 5 kilowatts and the installation complies with Rule 220B2.

225. Electric Railway Construction

A. Trolley-Contact Conductor Fastenings

All overhead trolley-contact conductors shall be supported and arranged so that the breaking of a single contact conductor fastening will not allow the trolley conductor live span wire, or current-carrying connection to come within 10 ft (3.0 m) (measured vertically) from the ground, or from any platform accessible to the general public.

Span-wire insulation for trolley-contact conductors shall comply with Rule 279B.

B. High-Voltage Contact Conductors

Trolley-contact conductors energized at more than 750 volts shall be suspended so as to minimize the possibility of a break, and in such a way that, if broken at one point, the conductor will not come within 12 ft (3.6 m) (measured vertically) of the ground, or any platform accessible to the public.

C. Third Rails

Third rails shall be protected by adequate guards composed of wood or other suitable insulating material.

EXCEPTION: This rule does not apply where third rails are on fenced right-of-way.

D. Prevention of Loss of Contact at Railroad Crossings at Grade

At crossings at grade with other railroads or other electrified railway systems, contact conductors shall be arranged as set forth in the following specifications, 1, 2, 3, 4, and 5, following whichever apply:

1. Where the crossing span exceeds 100 ft (30 m), catenary construction shall be used for overhead trolley-contact conductors.
2. When pole trolleys, using either wheels or sliding shoes, are used:
   a. The trolley-contact conductor shall be provided with live trolley guards of suitable construction; or
   b. The trolley-contact conductor should be at a uniform height above its own track throughout the crossing span and the next adjoining spans. Where it is not practical to maintain a uniform height, the change in height shall be made in a gradual manner.

   EXCEPTION: Rule 225D2 does not apply where the crossing is protected by signals or interlocking.
3. When pantograph type collectors are used, the contact conductor and track through the crossing should be maintained in a condition where the rocking of pantograph equipped cars or locomotives will not de-wire the pantograph. If this cannot be done, auxiliary contact conductors shall be installed. Wire height shall conform with Rule 225D2.
4. Where two electrified tracks cross:
   a. When the trolley-contact conductors are energized from different supply circuits, or from different phases of the same circuit, the trolley-conductor crossover shall be designed to insulate both conductors from each other. The design shall not permit either trolley collector to contact any conductor or part energized at a different voltage than at which it is designed to operate.
   b. Trolley-contact crossovers used to insulate trolley conductors of the same voltage but of different circuit sections shall be designed to prevent both sections being simultaneously contacted by the trolley collector.
5. When third rail construction is used, and the length of the third rail gap at the crossings is such that a car or locomotive stopping on the crossing can lose propulsion power, the crossing shall be protected by signals or interlocking.

E. Guards Under Bridges

Trolley guards of suitable construction shall be provided where the trolley-contact conductor is so located that a trolley pole leaving the conductor can make simultaneous contact between it and the bridge structure.

Section 23. Clearances

230. General

A. Application

This section covers all clearances, including climbing spaces, involving overhead supply and communications lines.

NOTE: The more than 70 years of historical development and specification of clearances in Rules 232, 233, and 234 were reviewed for consistency among themselves and with modern practice and were appropriately revised in both concept and content for the 1990 edition. See Appendix A.

B. Measurement of Clearance and Spacing

Unless otherwise stated, all clearances shall be measured from surface to surface and all spacings shall be measured center to center. For clearance measurement, live metallic hardware electrically connected to line conductors shall be considered a part of the line conductors. Metallic bases of potheads, surge arresters, and similar devices shall be considered a part of the supporting structure.

C. Supply Cables

For clearance purposes, supply cables, including splices and taps, conforming to any of the following requirements are permitted lesser clearances than open conductors of the same voltage. Cables should be capable of withstanding tests applied in accordance with an applicable standard.

1. Cables that are supported on or cabled together with an effectively grounded bare messenger or neutral, or with multiple concentric conductors where the associated neutral conductor(s) meet(s) the requirements of Rule 230E1 and where the cables also meet one of the following:
   a. cables of any voltage having an effectively grounded continuous metal sheath or shield, or
   b. cables designed to operate on a multigrounded system at 22 kV or less and having a semiconducting

insulating shield in combination with suitable metallic drainage.

2. Cables of any voltage, not included in Rule 230C1, covered with a continuous auxiliary semiconducting shield in combination with suitable metallic drainage and supported on and cabled together with an effectively grounded bare messenger.
3. Insulated, non-shielded cable operated at not over 5 kV phase-to-phase, or 2.9 kV phase-to-ground, supported on and cabled together with an effectively grounded bare messenger.

D. Covered Conductors

Covered conductors shall be considered bare conductors for all clearance requirements except that spacing between conductors of the same or different circuits, including grounded conductors, may be reduced below the requirements for open conductors when the conductors are owned, operated, or maintained by the same party and when the conductor covering provides sufficient dielectric strength to prevent a short circuit in case of momentary contact between conductors or between conductors and the grounded conductor. Intermediate spacers may be used to maintain conductor spacing and provide support.

E. Neutral Conductors

1. Neutral conductors which are effectively grounded throughout their length and associated with circuits of 0 to 22 kilovolts to ground may have the same clearances as guys and messengers.
2. All other neutral conductors of supply circuits shall have the same clearances as the phase conductors of the circuit with which they are associated.

F. Fiber Optic Cable

1. Cable defined as "fiber optic-supply" supported on a messenger that is effectively grounded throughout its length shall have the same clearance from communication facilities as required for a neutral conductor meeting Rule 230E1. Fiber optic-supply cables supported on or within other messengers or conductors shall have the same clearances from communication facilities required for such messengers or conductors.

EXCEPTION: Fiber optic-supply cables meeting Rule 224A3 may be considered as ordinary communication cables.

2. Cable defined as "fiber optic-communication" shall have the same clearance from supply facilities as required for a communication messenger.

G. Alternating and Direct Current Circuits

The rules of this section are applicable to both alternating and direct current circuits. For direct current circuits, the clearance requirements shall be the same as those for alternating current circuits having the same crest voltage to ground.

H. Constant-Current Circuits

The clearances for constant-current circuits shall be determined on the basis of their normal full-load voltage.

I. Maintenance of Clearances and Spacings

The clearances and spacing required shall be maintained at the values and under the conditions specified in Section 23.

231. Clearances of Supporting Structures from Other Objects

Supporting structures, support arms and equipment attached thereto, and braces shall have the following clearances from other objects. The clearance shall be measured between the nearest parts of the objects concerned.

A. From Fire Hydrants

Not Less Than 3 ft (900 mm).

RECOMMENDATION: Where conditions permit, a clearance of not less than 4 ft (1.20 m) is recommended.

B. From Streets, Roads, and Highways

1. Where there are curbs: supporting structures, support arms, or equipment attached thereto, up to 15 ft (4.6 m) above the road surface shall be located a sufficient distance from the street side of the curbs to avoid contact by ordinary vehicles using and located on the traveled way. In no case shall such distance be less than 6 in (150 mm).
2. Where there are no curbs, supporting structures should be located a sufficient distance from the roadway to avoid contact by ordinary vehicles using and located on the traveled way.
3. Location of overhead utility installations on highways with narrow rights-of-way or on urban streets with closely abutting improvements are special cases which

must be resolved in a manner consistent with the prevailing limitations and conditions.

4. Where a governmental authority exercising jurisdiction over structure location has issued a permit for, or otherwise approved, specific locations for supporting structures, that permit or approval shall govern.

C. From Railroad Tracks

Where railroad tracks are parallel or crossed by overhead lines, all portions of the supporting structures, support arms, anchor guys, and equipment attached thereto less than 22 ft (6.7 m) above the nearest track rail shall be located not less than 12 ft (3.6 m) from the nearest track rail. See Rule 234I.

EXCEPTION 1: A clearance of not less than 7 ft (2.13 m) may be allowed where the supporting structure is not the controlling obstruction, provided sufficient space for a driveway is left where cars are loaded or unloaded.

EXCEPTION 2: Supports for overhead trolley contact conductors may be located as near their own track rail as conditions require. If very close, however, permanent screens on cars will be necessary to protect passengers.

EXCEPTION 3: Where necessary to provide safe operating conditions which require an uninterrupted view of signals, signs, etc along tracks, the parties concerned shall cooperate in locating structures to provide the necessary clearance.

EXCEPTION 4: At industrial sidings, a clearance of not less than 7 ft (2.13 m) shall be permitted, provided sufficient space is left where cars can be loaded or unloaded.

232. Vertical Clearances of Wires, Conductors, Cables, and Equipment Above Ground, Roadway, Rail, or Water Surfaces

A. Application

The vertical clearances specified in Rule 232B1 apply under the following conductor temperature and loading conditions, whichever produces the largest final sag.

1. 120 °F (50 °C), no wind displacement.
2. The maximum conductor temperature for which the line is designed to operate, if greater than 120 °F (50 °C), with no wind displacement.
3. 32 °F (0 °C), no wind displacement, with radial thickness of ice, if any, specified in Rule 250B for the loading district concerned.

EXCEPTION: The conductor temperature and loading condition for trolley and electrified railroad contact conductors shall be 60 °F (15 °C), no wind displacement, final unloaded sag, or initial unloaded sag in cases where these facilities are maintained approximately at initial unloaded sags.

NOTE: The phase and neutral conductors of a supply line are normally considered separately when determining the sag of each due to temperature rise.

B. Clearance of Wires, Conductors, Cables, and Equipment Mounted on Supporting Structures

1. Clearance to Wires, Conductors, and Cables
   The vertical clearance of wires, conductors, and cables above ground in generally accessible places, roadway, rail, or water surfaces, shall be not less than that shown in Table 232-1.
2. Clearance to Unguarded Rigid Live Parts of Equipment
   The vertical clearance above ground or roadway surfaces for unguarded rigid live parts such as potheads, transformer bushings, surge arresters, and short lengths of supply conductors connected thereto, which are not subject to variation in sag, shall be not less than that shown in Table 232-2.
3. Clearance to Equipment Cases
   The vertical clearance of equipment cases above ground or roadway surfaces shall be not less than that shown in Table 232-2.
4. Street and Area Lighting
   a. All exposed ungrounded conductive parts of luminaires and their supports that are not insulated from current-carrying parts shall be maintained at not less than 20 in (500 mm) from the surface of their supporting structure.
      EXCEPTION 1: This may be reduced to 5 in (125 mm) if located on the side of the structure opposite the designated climbing space.
      EXCEPTION 2: This does not apply where the equipment is located at the top or other vertical portion of the structure that is not subject to climbing.
   b. Insulators, as specified in Rule 279A, should be inserted at least 8 ft (2.45 m) from the ground in metallic suspension ropes or chains supporting lighting units of series circuits.

C. Additional Clearances for Wires, Conductors, Cables, and Unguarded Rigid Live Parts of Equipment

Greater clearances than specified by Rule 232B shall be provided where required by Rule 232C1.

1. Voltages Exceeding 22 Kilovolts
   a. For voltages between 22 and 470 kilovolts, the clearance specified in Rule 232B1 (Table 232-1) or Rule 232B2 (Table 232-2) shall be increased at the rate of 0.4 in (10 mm) per kilovolt in excess of 22 kilovolts. For voltages exceeding 470 kilovolts, the clearance shall be determined by the method given in Rule 232D. All clearances for lines over 50 kilovolts shall be based on the maximum operating voltage.

      EXCEPTION: For voltages exceeding 98 kV alternating current to ground or 139 kV direct current to ground, clearances less than those required above are permitted for systems with known maximum switching surge factors (see Rule 232D).
   b. For voltages exceeding 50 kV, the additional clearance specified in Rule 232C1a shall be increased 3% for each 1000 ft (300 m) in excess of 3300 ft (1000 m) above mean sea level.
   c. For voltages exceeding 98 kV alternating current to ground, or 139 kV direct current to ground, either the clearances shall be increased or the electric field, or the effects thereof, shall be reduced by other means, as required, to limit the current due to electrostatic effects to 5.0 milliampere , rms, if the largest anticipated truck, vehicle, or equipment under the line were short circuited to ground. The size of the anticipated truck, vehicle, or equipment used to determine these clearances may be less than but need not be greater than that limited by Federal, State, or local regulations governing the area under the line. For this determination, the conductors shall be at a final unloaded sag at 120 °F (50 °C).

D. Alternate Clearances for Voltages Exceeding 98 Kilovolts Alternating Current to Ground or 139 Kilovolts Direct Current to Ground

The clearances specified in Rules 232B and 232C may be reduced for circuits with known switching surge factors, but shall be not less than the alternate clearance, which is computed by adding the reference height from Rule 232D2 to the electrical component of clearance from Rule 232D3.

1. Sag Conditions of Line Conductors
   The vertical clearance shall be maintained under the conductor temperature and loading condition given in Rule 232A.
2. Reference Heights
   The reference height shall be selected from Table 232-3.
3. Electrical Component of Clearance
   a. The electrical component (D) shall be computed using the following equations. Selected values of D are listed in Table 232-4.

$$D = 3.28 \left[\frac{V \cdot (PU) \cdot a}{500\,K}\right]^{1.667} bc \text{ (ft)}$$

$$D = 1.00 \left[\frac{V \cdot (PU) \cdot a}{500\,K}\right]^{1.667} bc \text{ (m)}$$

where

V = maximum alternating current crest operating voltage to ground or maximum direct current operating voltage to ground in kilovolts;

PU = maximum switching surge factor expressed in per-unit peak voltage to ground and defined as a switching surge level for circuit breakers corresponding to 98% probability that the maximum switching surge generated per breaker operation does not exceed this surge level, or the maximum anticipated switching surge level generated by other means, whichever is greater;

a = 1.15, the allowance for three standard deviations;

b = 1.03, the allowance for nonstandard atmospheric conditions;

c = 1.2, the margin of safety;

K = 1.15, the configuration factor for conductor-to-plane gap.

   b. The value of D shall be increased 3% for each 1000 ft (300 m) in excess of 1500 ft (450 m) above mean sea level.

c. Either the clearances shall be increased or the electric field, or the effects thereof, shall be reduced by other means, as required, to limit the current due to electrostatic effects to 5.0 milliamperes, rms, if the largest anticipated truck, vehicle, or equipment under the line were short circuited to ground. The size of the anticipated truck, vehicle, or equipment used to determine these clearances may be less than but need not be greater than that limited by Federal, State, or local regulations governing the area under the line. For this determination, the conductors shall be at a final unloaded sag at 120 °F (50 °C).

4. Limit

The alternate clearance shall be not less than the clearance given in Tables 232-1 or 232-2 computed for 98 kilovolts alternating current to ground in accordance with Rule 232C.

Table 232-1 Vertical Clearance of Wires, Conductors, and Cables Above Ground, Roadway, Rail, or Water Surfaces

(Voltages are phase-to-ground for effectively grounded circuits and those other circuits where all ground faults are cleared by promptly de-energizing the faulted section, both initially and following subsequent breaker operations. See the definition section for voltages of other systems.) **FT**

Nature of surface underneath wires, conductors, or cables	(11) Insulated communication conductors and cable; messengers; surge protection wires; grounded guys; neutral conductors meeting Rule 230E1; supply cables meeting Rule 230C1 (ft)	Non-insulated communication conductors; supply cables of 0 to 750 V meeting Rules 230C2 or 230C3 (ft)	Supply cables over 750 V meeting Rules 230C2 or 230C3; open supply conductors, 0 to 750 V (ft)	Open supply conductors, over 750 V to 22 kV (ft)	Trolley and electrified railroad contact conductors and associated span or messenger wires (1): 0 to 750 V to ground (ft)	Trolley and electrified railroad contact conductors and associated span or messenger wires (1): over 750 V to 22 kV to ground (ft)
Where wires, conductors, or cables cross over or overhang						
1. Track rails of railroads (except electrified railroads using overhead trolley conductors) (2) (16) (20)	23.5	24.0	24.5	26.5	22.0 (4)	22.0 (4)
2. Roads, streets, alleys; nonresidential driveways, parking lots, and other areas subject to truck traffic (21)	15.5 (13)	16.0 (13)	16.5	18.5	18.0 (5)	20.0 (5)
3. Residential driveways	15.5 (7) (13)	16.0 (7) (13)	16.5 (7)	18.5	18.0 (5)	20.0 (5)
4. Other land traversed by vehicles, such as cultivated, grazing, forest, orchard, etc	15.5	16.0	16.5	18.5	-	-
5. Spaces and ways subject to pedestrians or restricted traffic only (9)	9.5	12.0 (8)	12.5 (8)	14.5	16.0	18.0

6. Water areas not suitable for sailboating or where sailboating is prohibited (19)	14.0	14.5	15.0	17.0	-	-
7. Water areas suitable for sailboating including lakes, ponds, reservoirs, tidal waters, rivers, streams, and canals with an unobstructed surface area of: (17) (18) (19)						
(a) Less than 20 acres	17.5	18.0	18.5	20.5	-	-
(b) 20 to 200 acres	25.5	26.0	26.5	28.5	-	-
(c) Over 200 to 2000 acres	31.5	32.0	32.5	34.5	-	-
(d) Over 2000 acres	37.5	38.0	38.5	40.5	-	-
8. Public or private land and water areas posted for rigging or launching sailboats	Clearance above ground shall be 5 ft greater than in 7 above, for the type of water areas served by the launching site					
Where wires, conductors, or cables run along and within the limits of highways or other road rights-of-way but do not overhang the roadway						
9. Roads, streets, or alleys	15.5 (13) (24)	16.0 (13)	16.5	18.5	18.0 (5)	20.0 (5)
10. Roads in rural districts where it is unlikely that vehicles will be crossing under the line	13.5 (10) (12)	14.0 (10)	14.5 (10)	16.5	18.0 (5)	20.0 (5)

(1) Where subways, tunnels, or bridges require it, less clearances above ground or rails than required by Table 232-1 may be used locally. The trolley and electrified railroad contact conductor should be graded very gradually from the regular construction down to the reduced elevation.

(2) For wire, conductors, or cables crossing over mine, logging, and similar railways which handle only cars lower than standard freight cars, the clearance may be reduced by an amount equal to the difference in height between the highest loaded car handled and 22 ft, but the clearances shall not be reduced below that required for street crossings.

(3) This footnote not used in this edition.

(4) In communities where 21 ft has been established, this clearance may be continued if carefully maintained. The elevation of the contact conductor should be the same in the crossing and next adjacent spans. (See Rule 225D2 for conditions which must be met where uniform height above rail is impractical.)

(5) In communities where 16 ft has been established for trolley and

electrified railroad contact conductors 0 to 750 V to ground, or 18 ft for trolley and electrified railroad contact conductors exceeding 750 V, or where local conditions make it impractical to obtain the clearance given in the table, these reduced clearances may be used if carefully maintained.

⑥ This footnote not used in this edition.

⑦ Where the height of attachment to a building or other installation does not permit service drops to meet these values, the clearances may be reduced to the following:

		(feet)
(a)	Insulated supply service drops limited to 300 V to ground	12.5
(b)	Insulated drip loops of supply service drops limited to 300 V to ground	10.5
(c)	Supply service drops limited to 150 V to ground and meeting Rules 230C1 or 230C3	12.0
(d)	Drip loops only of service drops limited to 150 V to ground and meeting Rules 230C1 or 230C3	10.0
(e)	Insulated communication service drops.	11.5

⑧ Where the height of attachment to a building or other installation does not permit service drops to meet these values, the clearances may be reduced to the following:

		(feet)
(a)	Insulated supply service drops limited to 300 V to ground	10.5
(b)	Insulated drip loops of supply service drops limited to 300 V to ground	10.5
(c)	Supply service drops limited to 150 V to ground and meeting Rules 230C1 or 230C3	10.0
(d)	Drip loops only of supply service drops limited to 150 V to ground and meeting Rules 230C1 or 230C3.	10.0

⑨ Spaces and ways subject to pedestrians or restricted traffic only are those areas where equestrians, vehicles, or other mobile units, exceeding 8 ft in height, are prohibited by regulation or permanent terrain configurations or are otherwise not normally encountered or not reasonably anticipated.

⑩ Where a supply or communication line along a road is located relative to fences, ditches, embankments, etc, so that the ground under the line would not be expected to be travelled except by pedestrians, this clearance may be reduced to the following values:

		(feet)
(a)	Insulated communication conductor and communication cables	9.5
(b)	Conductors of other communication circuits	9.5
(c)	Supply cables of any voltage meeting Rule 230C1 and supply cables limited to 150 V to ground meeting Rules 230C2 or 230C3	9.5
(d)	Insulated supply conductors limited to 300 V to ground	12.5
(e)	Guys	9.5

⑪ No clearance from ground is required for anchor guys not crossing tracks, rails, streets, driveways, roads, or pathways.

⑫ This clearance may be reduced to 13 ft for communication conductors and guys.

⑬ Where this construction crosses over or runs along alleys, driveways, or parking lots, this clearance may be reduced to 15 ft.

⑭ This footnote not used in this edition.

⑮ This footnote not used in this edition.

⑯ Adjacent to tunnels and overhead bridges which restrict the height of loaded rail cars to less than 22 ft, these clearances may be reduced by the difference between the highest loaded rail car handled and 22 ft, if mutually agreed to by the parties at interest.

⑰ For controlled impoundments, the surface area and corresponding clearances shall be based upon the design high water level. For other waters, the surface area shall be that enclosed by its annual high water mark, and clearances shall be based on the normal flood level. The clearance over rivers, streams, and canals shall be based upon the largest surface area of any 1 mi long segment which includes the crossing. The clearance over a canal, river, or stream normally used to provide access for sailboats to a larger body of water shall be the same as that required for the larger body of water.

(18) Where an overwater obstruction restricts vessel height to less than the applicable reference height given in Table 232-3, the required clearance may be reduced by the difference between the reference height and the overwater obstruction height, except that the reduced clearance shall be not less than that required for the surface area on the line-crossing side of the obstruction.

(19) Where the US Army Corps of Engineers, or the State, or surrogate thereof has issued a crossing permit, clearances of that permit shall govern.

(20) See Rule 234I for the required horizontal and diagonal clearances to rail cars.

(21) For the purpose of this rule, trucks are defined as any vehicle exceeding 8 ft in height. Areas not subject to truck traffic are areas where truck traffic is not normally encountered or not reasonably anticipated.

(22) This footnote not used in this edition.

(23) This footnote not used in this edition.

(24) Communication cables and conductors may have a clearance of 15 ft where poles are back of curbs or other deterrents to vehicular traffic.

Table 232-1 Vertical Clearance of Wires, Conductors, and Cables Above Ground, Roadway, Rail, or Water Surfaces

(Voltages are phase-to-ground for effectively grounded circuits and those other circuits where all ground faults are cleared by promptly de-energizing the faulted section, both initially and following subsequent breaker operations. See the definition section for voltages of other systems.) **M**

Nature of surface underneath wires, conductors, or cables	(11) Insulated communication conductors and cable; messengers; surge protection wires; grounded guys; neutral conductors meeting Rule 230E1; supply cables meeting Rule 230C1 (m)	Non-insulated communication conductors; supply cables of 0 to 750 V meeting Rules 230C2 or 230C3 (m)	Supply cables over 750 V meeting Rules 230C2 or 230C3; open supply conductors, 0 to 750 V (m)	Open supply conductors, over 750 V to 22 kV (m)	Trolley and electrified railroad contact conductors and associated span or messenger wires (1): 0 to 750 V to ground (m)	Trolley and electrified railroad contact conductors and associated span or messenger wires (1): over 750 V to 22 kV to ground (m)
Where wires, conductors, or cables cross over or overhang						
1. Track rails of railroads (except electrified railroads using overhead trolley conductors) (2) (16) (20)	7.2	7.3	7.5	8.1	6.7 (4)	6.7 (4)
2. Roads, streets, alleys; nonresidential driveways, parking lots, and other areas subject to truck traffic (21)	4.7 (13)	4.9 (13)	5.0	5.6	55 (5)	6.1 (5)
3. Residential driveways	4.7 (7) (13)	4.9 (7) (13)	5.0 (7)	5.6	5.5 (5)	6.1 (5)
4. Other land traversed by vehicles, such as cultivated, grazing, forest, orchard, etc	4.7	4.9	5.0	5.6	-	-
5. Spaces and ways subject to pedestrians or restricted traffic only (9)	2.90	3.6 (8)	3.8 (8)	4.4	4.9	5.5

6. Water areas not suitable for sailboating or where sailboating is prohibited (19)	4.3	4.4	4.6	5.2	-	-
7. Water areas suitable for sailboating including lakes, ponds, reservoirs, tidal waters, rivers, streams, and canals with an unobstructed surface area of: (17) (18) (19)						
(a) **Less than 8 ha**	5.3	5.5	5.6	6.2	-	-
(b) **8 to 80 ha**	7.8	7.9	8.1	8.7	-	-
(c) **80 to 800 ha**	9.6	9.8	9.9	10.5	-	-
(d) **Over 800 ha**	11.4	11.6	11.7	12.3	-	-
8. Public or private land and water areas posted for rigging or launching sailboats	Clearance above ground shall be 1.5 m greater than in 7 above, for the type of water areas served by the launching site					
Where wires, conductors, or cables run along and within the limits of highways or other road rights-of-way but do not overhang the roadway						
9. Roads, streets, or alleys	4.7 (13) (24)	4.9 (13)	5.0	5.6	5.5. (5)	6.1 (5)
10. Roads in rural districts where it is unlikely that vehicles will be crossing under the line	4.1 (10) (12)	4.3 (10)	4.4 (10)	5.0	5.5 (5)	6.1 (5)

(1) Where subways, tunnels, or bridges require it, less clearances above ground or rails than required by Table 232-1 may be used locally. The trolley and electrified railroad contact conductor should be graded very gradually from the regular construction down to the reduced elevation.

(2) For wire, conductors, or cables crossing over mine, logging, and similar railways which handle only cars lower than standard freight cars, the clearance may be reduced by an amount equal to the difference in height between the highest loaded car handled and 6.7 m, but the clearances shall not be reduced below that required for street crossings.

(3) This footnote not used in this edition.

(4) In communities where 6.4 m has been established, this clearance may be continued if carefully maintained. The elevation of the contact conductor should be the same in the crossing and next adjacent spans. (See Rule 225D2 for conditions which must be met where uniform height above rail is impractical.)

(5) In communities where 4.9 m has been established for trolley and

electrified railroad contact conductors 0 to 750 V to ground, or 5.5 m for trolley and electrified railroad contact conductors exceeding 750 V, or where local conditions make it impractical to obtain the clearance given in the table, these reduced clearances may be used if carefully maintained.

⑥ This footnote not used in this edition.

⑦ Where the height of attachment to a building or other installation does not permit service drops to meet these values, the clearances may be reduced to the following:

		(meters)
(a)	Insulated supply service drops limited to 300 V to ground	3.8
(b)	Insulated drip loops of supply service drops limited to 300 V to ground	3.2
(c)	Supply service drops limited to 150 V to ground and meeting Rules 230C1 or 230C3	3.6
(d)	Drip loops only of service drops limited to 150 V to ground and meeting Rules 230C1 or 230C3	3.0
(e)	Insulated communication service drops.	3.5

⑧ Where the height of attachment to a building or other installation does not permit service drops to meet these values, the clearances may be reduced to the following:

		(meters)
(a)	Insulated supply service drops limited to 300 V to ground	3.2
(b)	Insulated drip loops of supply service drops limited to 300 V to ground	3.2
(c)	Supply service drops limited to 150 V to ground and meeting Rules 230C1 or 230C3	3.0
(d)	Drip loops only of supply service drops limited to 150 V to ground and meeting Rules 230C1 or 230C3.	3.0

⑨ Spaces and ways subject to pedestrians or restricted traffic only are those areas where equestrians, vehicles, or other mobile units, exceeding 2.45 m in height, are prohibited by regulation or permanent terrain configurations or are otherwise not normally encountered or not reasonably anticipated.

⑩ Where a supply or communication line along a road is located relative to fences, ditches, embankments, etc, so that the ground under the line would not be expected to be travelled except by pedestrians, this clearance may be reduced to the following values:

		(meters)
(a)	Insulated communication conductor and communication cables	2.90
(b)	Conductors of other communication circuits	2.90
(c)	Supply cables of any voltage meeting Rule 230C1 and supply cables limited to 150 V to ground meeting Rules 230C2 or 230C3	2.90
(d)	Insulated supply conductors limited to 300 V to ground	3.8
(e)	Guys	2.90

⑪ No clearance from ground is required for anchor guys not crossing tracks, rails, streets, driveways, roads, or pathways.

⑫ This clearance may be reduced to 4.0 m for communication conductors and guys.

⑬ Where this construction crosses over or runs along alleys, driveways, or parking lots, this clearance may be reduced to 4.6 m.

⑭ This footnote not used in this edition.

⑮ This footnote not used in this edition.

⑯ Adjacent to tunnels and overhead bridges which restrict the height of loaded rail cars to less than 6.7 m, these clearances may be reduced by the difference between the highest loaded rail car handled and 6.7 m, if mutually agreed to by the parties at interest.

⑰ For controlled impoundments, the surface area and corresponding clearances shall be based upon the design high water level. For other waters, the surface area shall be that enclosed by its annual high water mark, and clearances shall be based on the normal flood level. The clearance over rivers, streams, and canals shall be based upon the largest surface area of any 1.6 km long segment which includes the crossing. The clearance over a canal, river, or stream normally used to provide access for sailboats to a larger body of water shall be the same as that required for the larger body of water.

(18) Where an overwater obstruction restricts vessel height to less than the applicable reference height given in Table 232-3, the required clearance may be reduced by the difference between the reference height and the overwater obstruction height, except that the reduced clearance shall be not less than that required for the surface area on the line-crossing side of the obstruction.

(19) Where the US Army Corps of Engineers, or the State, or surrogate thereof has issued a crossing permit, clearances of that permit shall govern.

(20) See Rule 234I for the required horizontal and diagonal clearances to rail cars.

(21) For the purpose of this rule, trucks are defined as any vehicle exceeding 2.45 m in height. Areas not subject to truck traffic are areas where truck traffic is not normally encountered or not reasonably anticipated.

(22) This footnote not used in this edition.

(23) This footnote not used in this edition.

(24) Communication cables and conductors may have a clearance of 4.6 m where poles are back of curbs or other deterrents to vehicular traffic.

Table 232-2 Vertical Clearance of Equipment Cases and Unguarded Rigid Live Parts Above Ground or Roadway Surfaces

(Voltages are phase-to-ground for effectively grounded circuits and those other circuits where all ground faults are cleared by promptly de-energizing the faulted section, both initially and following subsequent breaker operations. See the definition section for voltages of other systems.) **FT**

Nature of surface below:	Effectively grounded equipment cases (6) (ft)	Unguarded rigid live parts of 0 to 750 V and ungrounded cases that contain equipment connected to circuits of not more than 750 V (ft)	Unguarded rigid live parts of over 750 V to 22 kV and ungrounded cases that contain equipment connected to circuits of over 750 V to 22 kV (ft)
1. Where rigid parts overhang:			
a. Roads, streets, alleys; nonresidential driveways; parking lots and other areas subject to truck traffic (4)	15.0	16.0	18.0
b. Residential driveways	15.0	16.0 (1)	18.0
c. Other land traversed by vehicles such as cultivated land, grazing land, forest, orchard, etc.	15.0	16.0	18.0
d. Spaces and ways subject to pedestrians or restricted traffic only (5)	11.0 (7)	12.0 (1b)	14.0
2. Where rigid parts are along and within the limits of highways or other road rights-of-way but do not overhang the roadway:			

a.	Roads, streets, and alleys	15.0	16.0	18.0
b.	Roads in rural districts where it is unlikely that vehicles will be crossing under the line	13.0 ⑦	14.0 ②	16.0

① This clearance may be reduced to the following values:

		(feet)
a.	Insulated live parts limited to 300 V to ground	12
b.	Insulated live parts limited to 150 V to ground and drip loops of service drop conductors limited to 150 V to ground and meeting Rules 230C2 or 230C3.	10

② Where a supply line along a road is limited to 300 V to ground and is located relative to fences, ditches, embankments, etc, so that the ground under the line would not be expected to be traveled except by pedestrians, this clearance may be reduced to 12 ft.

③ This footnote not used in this edition.

④ For the purpose of this rule, trucks are defined as any vehicle exceeding 8 ft in height. Areas not subject to truck traffic are areas where truck traffic is not normally encountered or not reasonably anticipated.

⑤ Spaces and ways subject to pedestrians or restricted traffic only are those areas where equestrians, vehicles, or other mobile units, exceeding 8 ft in height, are prohibited by regulation or permanent terrain configurations or are otherwise not normally encountered or not reasonably anticipated.

⑥ The bottom of the housing of traffic control signals suspended over the traveled portion of the roadway shall be not less than 15 ft nor more than 19 ft above the grade at the center of the roadway.

⑦ Effectively grounded equipment cases such as fire alarm boxes, traffic control boxes, or meters may be mounted over a walkway at a lower level for accessibility provided such equipment does not unduly obstruct the walkway.

Table 232-2 Vertical Clearance of Equipment Cases and Unguarded Rigid Live Parts Above Ground or Roadway Surfaces

(Voltages are phase-to-ground for effectively grounded circuits and those other circuits where all ground faults are cleared by promptly de-energizing the faulted section, both initially and following subsequent breaker operations. See the definition section for voltages of other systems.) **M**

Nature of surface below:	Effectively grounded equipment cases ⑥ (m)	Unguarded rigid live parts of 0 to 750 V and ungrounded cases that contain equipment connected to circuits of not more than 750 V (m)	Unguarded rigid live parts of over 750 V to 22 kV and ungrounded cases that contain equipment connected to circuits of over 750 V to 22 kV (m)
1. Where rigid parts overhang:			
a. Roads, streets, alleys; nonresidential driveways; parking lots and other areas subject to truck traffic ④	4.6	4.9	5.5
b. Residential driveways	4.6	4.9 ①	5.5
c. Other land traversed by vehicles such as cultivated land, grazing land, forest, orchard, etc.	4.6	4.9	5.5
d. Spaces and ways subject to pedestrians or restricted traffic only ⑤	3.4 ⑦	3.6 (1b)	4.3
2. Where rigid parts are along and within the limits of highways or other road rights-of-way but do not overhang the roadway:			

a.	Roads, streets, and alleys	4.6	4.9	5.5
b.	Roads in rural districts where it is unlikely that vehicles will be crossing under the line	4.0 ⑦	4.3 ②	4.9

① This clearance may be reduced to the following values:

		(meters)
a.	Insulated live parts limited to 300 V to ground	3.6
b.	Insulated live parts limited to 150 V to ground and drip loops of service drop conductors limited to 150 V to ground and meeting Rules 230C2 or 230C3.	3.0

② Where a supply line along a road is limited to 300 V to ground and is located relative to fences, ditches, embankments, etc, so that the ground under the line would not be expected to be traveled except by pedestrians, this clearance may be reduced to 3.6 m.

③ This footnote not used in this edition.

④ For the purpose of this rule, trucks are defined as any vehicle exceeding 2.45 m in height. Areas not subject to truck traffic are areas where truck traffic is not normally encountered or not reasonably anticipated.

⑤ Spaces and ways subject to pedestrians or restricted traffic only are those areas where equestrians, vehicles, or other mobile units, exceeding 2.45 m in height, are prohibited by regulation or permanent terrain configurations or are otherwise not normally encountered or not reasonably anticipated.

⑥ The bottom of the housing of traffic control signals suspended over the traveled portion of the roadway shall be not less than 4.6 m nor more than 5.8 m above the grade at the center of the roadway.

⑦ Effectively grounded equipment cases such as fire alarm boxes, traffic control boxes, or meters may be mounted over a walkway at a lower level for accessibility provided such equipment does not unduly obstruct the walkway.

Table 232-3
Reference Heights

Nature of surface underneath lines	(ft)	(m)
a. Track rails of railroads (except electrified railroads using overhead trolley conductors) ①	22	6.7
b. Streets, alleys, roads, driveways, and parking lots	14	4.3
c. Spaces and ways subject to pedestrians or restricted traffic only ②	10	3.0
d. Other land, such as cultivated, grazing, forest, or orchard, which is traversed by vehicles	14	4.3
e. Water areas not suitable for sailboating or where sailboating is prohibited	12.5	3.8
f. Water areas suitable for sailboating including lakes, ponds, reservoirs, tidal waters, rivers, streams, and canals with unobstructed surface area ③ ④		
(1) less than 20 acres (8 ha)	16	4.9
(2) 20 to 200 acres (8 to 80 ha)	24	7.3
(3) over 200 to 2000 acres (80 to 800 ha)	30	9.0
(4) over 2000 acres (800 ha)	36	11.0
g. In public or private land and water areas posted for rigging or launching sailboats, the reference height shall be 5 ft (1.5 m) greater than in f above, for the type of water areas serviced by the launching site		

① See Rule 234I for the required horizontal and diagonal clearances to rail cars.

② Spaces and ways subject to pedestrians or restricted traffic only are those areas where equestrians, vehicles, or other mobile units, exceeding 8 ft (2.45 m) in height, are prohibited by regulation or permanent terrain configurations or are otherwise not normally encountered or not reasonably anticipated.

③ For controlled impoundments, the surface area and corresponding clearances shall be based upon the design high water level. For other waters, the surface area shall be that enclosed by its annual high water mark, and clearances shall be based on the normal flood level. The clearances over rivers, streams, and canals shall be based upon the largest surface area of any 1 mile long (1600 m) segment which includes the crossing. The clearance over a canal or similar waterway providing access for sailboats to a larger body of water shall be the same as that required for the larger body of water.

④ Where an overwater obstruction restricts vessel height to less than the applicable reference height, the required clearance may be reduced by the difference between the reference height and the overwater obstruction height, except that the reduced clearance shall be not less than that required for the surface area on the line-crossing side of the obstruction.

Table 232-4 Electrical Component of Clearance in Rule 232D3a

(Add 3% for each 1000 ft (300 m) in excess of 1500 ft (450 m) above mean sea level. Increase clearance to limit electrostatic effects in accordance with Rule 232D3c.)

Maximum operating voltage phase-to-phase (kV)	Switching surge factor (per unit)	Switching surge (kV)	Electrical component of clearance (ft)	(m)
242	3.97 or less	785 or less	① 8.6	2.60
362	2.65 or less	785 or less	① 8.6	2.60
550	1.75 or less	785 or less	① 8.6	2.60
	1.90	853	9.9	3.0
	2.00	898	10.8	3.3
	2.20	988	12.7	3.9
	2.40	1079	14.6	4.4
	2.60	1168	16.7	5.1
800	1.60	1045	13.9	4.2
	1.80	1176	16.9	5.2
	2.00	1306	20.1	6.1
	2.10 or more	1372 or more	②21.8	6.6

① Limited by Rule 232D4.

② Limited by Rules 232A and 232B.

233. Clearances Between Wires, Conductors, and Cables Carried on Different Supporting Structures

A. General

Crossings should be made on a common supporting structure, where practical. In other cases, the clearance between any two crossing or adjacent wires, conductors, or cables carried on different supporting structures shall be not less at any location in the spans than that required by Rules 233B and 233C. The clearance shall be not less than that required by application of a clearance envelope developed under Rule 233A2 to the positions on or within conductor movement envelopes developed under Rule 233A1 at which the two wires, conductors, or cables would be closest together. For purposes of this determination, the relevant positions of the wires, conductors, or cables on or within their respective conductor movement envelopes are those which can occur when (1) both are simultaneously subjected to the same ambient air temperature and wind loading conditions and (2) each is subjected individually to the full range of its icing conditions and applicable design electrical loading.

Figure 233-1 is a graphical illustration of the application of Rule 233A. Alternate methods that assure compliance with these rules may be used.

1. Conductor Movement Envelope
   a. Development
      The conductor movement envelope shall be developed from the locus of the most displaced conductor positions defined below and shown in Fig 233-2.
      (1) 60 °F (15 °C), no wind displacement, at both initial unloaded and final unloaded sag (conductor positions A and C).
      (2) With the wire, conductor, or cable displaced from rest by a six pounds per square foot (290 Pa) wind at both initial and final sag at 60 °F (15 °C). This may be reduced to 4 pounds per square foot (190 Pa) wind in areas sheltered by buildings, terrain, or other obstacles. The displacement of the wire, conductor, or cable shall include deflection of suspension insulators

and flexible structures (conductor positions B and D).

(3) Final sag at one of the following loading conditions, whichever produces the largest sag (conductor position E):

(a) 120 °F (50 °C), no wind displacement, or

(b) The maximum conductor temperature for which the line is designed to operate, if greater than 120 °F (50 °C), with no wind displacement, or

(c) 32 °F (0 °C), no wind displacement, with radial thickness of ice, if any, specified in Rule 250B for the loading district concerned.

b. Sag Increase

No sag increase for either high operating temperatures or ice loading is required for trolley and electrified railroad contact conductors. Rule 233A1a(3) does not apply to these conductors.

2. Clearance Envelope

The clearance envelope shown in Fig 233-3 shall be determined by the horizontal clearance (H) required by Rule 233B and the vertical clearance (V) required by Rule 233C.

B. Horizontal Clearance

1. Clearance Requirements

The horizontal clearance between crossing or adjacent wires, conductors or cables carried on different supporting structures shall be not less than 5 ft (1.50 m). For voltages between the wires, conductors, or cables exceeding 129 kV, additional clearance of 0.4 in (10 mm) per kV over 129 kV shall be provided.

EXCEPTION: The horizontal clearance between anchor guys of different supporting structures may be reduced to 6 in (150 mm) and may be reduced to 2 ft (600 mm) between other guys, span wires and neutral conductors meeting Rule 230E1.

2. Alternate Clearances for Voltages Exceeding 98 kV Alternating Current to Ground or 139 kV Direct Current to Ground

The clearances specified in Rule 233B1 may be reduced for circuits with known switching surge factors, but shall be not less than the alternate clearance derived from the computations required in Rules 235B3a and 235B3b.

C. Vertical Clearance

1. Clearance Requirements

The vertical clearance between any crossing or adjacent wires, conductors, or cables carried on different supporting structures shall be not less than that shown in Table 233-1.

EXCEPTION: No vertical clearance is required between wires, conductors, or cables that are electrically interconnected at the crossing.

2. Voltages Exceeding 22 Kilovolts

a. The clearance given in Table 233-1 shall be increased by the sum of the following: For the upper level conductors between 22 and 470 kilovolts, the clearance shall be increased at the rate of 0.4 in (10 mm) per kilovolt in excess of 22 kilovolts. For the lower level conductors exceeding 22 kilovolts, the additional clearance shall be computed at the same rate. For voltages exceeding 470 kilovolts, the clearance shall be determined by the method given in Rule 233C3. The additional clearance shall be computed using the maximum operating voltage if above 50 kilovolts and nominal voltage if below 50 kilovolts.

EXCEPTION 1: Where a clearance value is given for the 22 kV to 50 kV range, the voltage adder of Rule 233C2a applies to voltage in excess of 50 kV.

EXCEPTION 2: For voltages exceeding 98 kilovolts alternating current to ground or 139 kilovolts direct current to ground, clearances less than those required above are permitted for systems with known switching surge factors. (See Rule 233C3.)

b. For voltages exceeding 50 kV, the additional clearance specified in Rule 232C2a shall be increased 3% for each 1000 ft (300 m) in excess of 3300 ft (1000 m) above mean sea level.

3. Alternate Clearances for Voltage Exceeding 98 Kilovolts Alternating Current to Ground or 139 Kilovolts Direct Current to Ground

The clearances specified in Rules 233C1 and 233C2 may be reduced where the higher voltage circuit has a known switching surge factor, but shall be not less than the alternate clearance, which is computed by adding the reference height from Rule 233C3a to the electrical component of clearance from Rule 233C3b. For these computations, communication conductors and cables, guys, messengers, neutral conductors meeting Rule 230E1, and supply cables meeting Rule 230C1 shall be considered at zero voltage.

a. Reference Heights
   The reference height shall be selected from Table 233-3.

b. Electrical Component of Clearance
   (1) The electrical component (D) shall be computed using the following equations. Selected values of D are listed in Table 233-2.

$$D = 3.28 \left[\frac{[V_H \cdot (PU) + V_L]\, a}{500\, K}\right]^{1.667} bc \quad \text{(ft)}$$

$$D = 1.00 \left[\frac{[V_H \cdot (PU) + V_L]\, a}{500\, K}\right]^{1.667} bc \quad \text{(m)}$$

where

V_H = higher voltage circuit maximum alternating current crest operating voltage to ground or maximum direct current operating voltage to ground in kilovolts;

V_L = lower voltage circuit maximum alternating current crest operating voltage to ground or maximum direct current operating voltage to ground in kilovolts;

PU = higher voltage circuit maximum switching surge factor expressed in per-unit peak voltage to ground and defined as a switching surge level for circuit breakers corresponding to 98% probability that the maximum switching surge generated per breaker operation does not exceed this surge level, or the maximum anticipated switching surge level generated by other means, whichever is greater;

a = 1.15, the allowance for three standard deviations;

b = 1.03, the allowance for nonstandard atmospheric conditions;

c = 1.2, the margin of safety;

K = 1.4, the configuration factor for conductor to-conductor gap;

(2) The value of D calculated by Rule 233C3b(1) shall be increased 3% for each 1000 feet (300 mm) in excess of 1500 feet (450 mm) above mean sea level.

c. Limit

The alternate clearance shall be not less than the clearance required by Rules 233C1 and 233C2 with the lower voltage circuit at ground potential.

Fig 233-1
Use of Clearance Envelope and Conductor Movement Envelopes to Determine Applicable Clearance

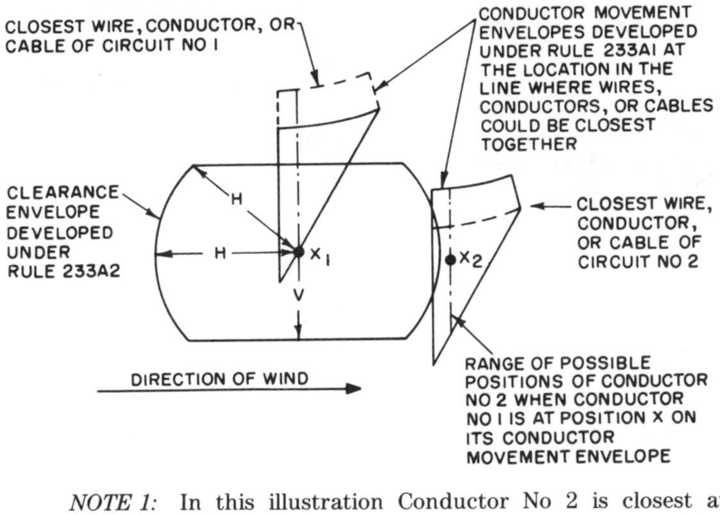

NOTE 1: In this illustration Conductor No 2 is closest at position X_2 to Conductor No 1, where the latter is at position X_1.

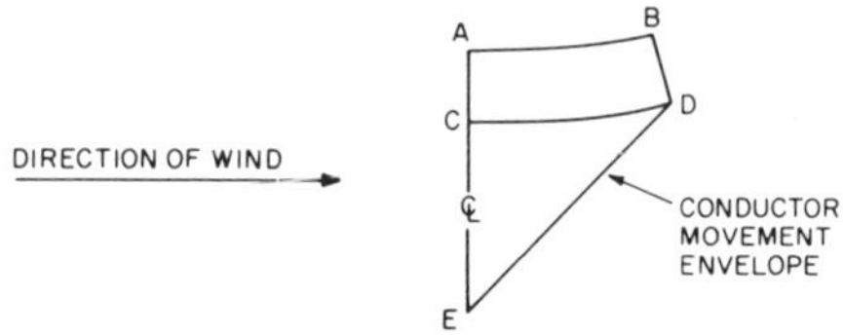

Point	Conductor Temperature	Sag	Ice Loading	Wind Displacement ①
A	60 °F	initial	none	none
B	60 °F	initial	none	6 lb per sq ft②
C	60 °F	final	none	none
D	60 °F	final	none	6 lb per sq ft②
$E_1$③④	The greater of 120 °F or maximum operating	final	none	none
$E_2$③④	32 °F	final	as applicable	none

① The direction of the wind shall be that which produces the minimum distance between conductors. The displacement of the wires, conductors, or cables includes the deflection of suspension insulators and flexible structures.

② Wind loading may be reduced to 4 lb per sq ft in areas sheltered by buildings, terrain, or other obstacles.

③ Point E shall be determined by whichever of the conditions described under E_1 and E_2 produces the greatest sag.

④ Line D–E shall be considered to be straight unless the actual concavity characteristics are known.

Fig 233-2 **FT**
Conductor Movement Envelope

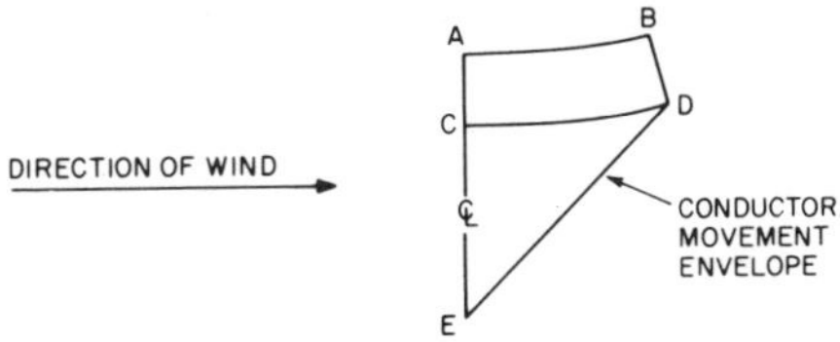

Point	Conductor Temperature	Sag	Ice Loading	Wind Displacement ①
A	15 °C	initial	none	none
B	15 °C	initial	none	290 Pa ②
C	15 °C	final	none	none
D	15 °C	final	none	290 Pa ②
E_1 ③④	The greater of 50 °C or maximum operating	final	none	none
E_2 ③④	0 °C	final	as applicable	none

① The direction of the wind shall be that which produces the minimum distance between conductors. The displacement of the wires, conductors, or cables includes the deflection of suspension insulators and flexible structures.

② Wind loading may be reduced to 190 Pa in areas sheltered by buildings, terrain, or other obstacles.

③ Point E shall be determined by whichever of the conditions described under E_1 and E_2 produces the greatest sag.

④ Line D–E shall be considered to be straight unless the actual concavity characteristics are known.

Fig 233-2 **M**

Conductor Movement Envelope

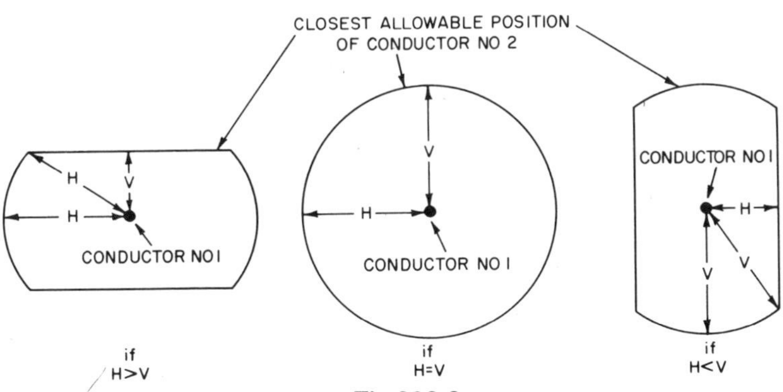

Fig 233-3

Clearance Envelope

Table 233-1 Vertical Clearance between Wires, Conductors, and Cables Carried on Different Supporting Structures

FT

(Voltages are phase to ground for effectively grounded circuits and those other circuits where all ground faults are cleared by promptly de-energizing the faulted section, both initially and following subsequent breaker operations. See the definitions section for voltages of other systems.)

Upper level / Lower level	Communication conductors and cables, and messengers (ft)	Guys, span wires, neutral conductors meeting Rule 230E1, and surge protection wires (ft)	Supply cables meeting Rule 230C1, and supply cables of 0 to 750 V meeting Rule 230C2 or 230C3 (ft)	Open supply conductors 0 to 750 V, supply cables over 750 V meeting Rule 230C2 or 230C3 (ft)	Open supply conductors over 750 V to 22 kV (ft)
1. Communication conductors and cables, and messengers	2 (2)	2	2	4 (8)	6 (5)
2. Supply cables meeting Rule 230C1, and supply cables of 0 to 750 V meeting Rules 230C2 or 230C3	2	2	2	2	2

3. Open supply conductors, 0 to 750 V; supply cables over 750 V meeting Rules 230C2 or 230C3	4 ⑨	2	4	2	2
4. Open supply conductors					
750 V to 22 kV	6 ⑤⑨	4	4 ⑨	4 ⑨	2
over 22 kV to 50 kV	6 ⑨	4	6 ⑨	6 ⑨	4 ⑨
5. Trolley and electrified railroad contact conductors and associated span and messenger wires	4 ③	4 ③	4 ③	4 ③④	6
6. Guys ⑦, span wires, neutral conductors meeting Rule 230E1, and surge protection wires	2 ②	2 ①②	2 ②	2	4

① This clearance may be reduced where both guys are electrically interconnected.

② The clearance of communication conductors and their guy, span, and messenger wires from each other in locations where no other classes of conductors are involved may be reduced by mutual consent of the parties concerned, subject to the approval of the regulatory body having jurisdiction, except for fire-alarm conductors and conductors used in the operation of railroads, or where one set of conductors is for public use and the other used in the operation of supply systems.

③ Trolley and electrified railroad contact conductors of more than 750 V should have at least 6 ft clearance. This clearance should also be provided over lower voltage trolley and electrified railroad contact conductors unless the crossover conductors are beyond reach of a

(continued on next page)

Footnotes for Table 233-1 *(Continued)*

trolley pole leaving the trolley-contact conductor or are suitably protected against damage from trolley poles leaving the trolley-contact conductor.

④ Trolley and electrified railroad feeders are exempt from this clearance requirement for contact conductors if they are of the same nominal voltage and of the same system.

⑤ This clearance may be reduced to 4 ft where supply conductors of 750 V to 8.7 kV cross a communication line more than 6 ft horizontally from a communication structure.

⑥ This footnote not used in this edition.

⑦ These clearances may be reduced by not more than 25% to a guy insulator, provided that full clearance is maintained to its metallic end fittings and the guy wires. The clearance to an insulated section of a guy between two insulators may be reduced by not more than 25% provided that full clearance is maintained to the uninsulated portion of the guy.

⑧ This clearance may be reduced to 2 ft for supply service drops.

⑨ In general, this type of crossing is not recommended.

M

Table 233-1 Vertical Clearance between Wires, Conductors, and Cables Carried on Different Supporting Structures

(Voltages are phase to ground for effectively grounded circuits and those other circuits where all ground faults are cleared by promptly de-energizing the faulted section, both initially and following subsequent breaker operations. See the definitions section for voltages of other systems.)

Upper level / Lower level	Communication conductors and cables, and messengers (m)	Guys, span wires, neutral conductors meeting Rule 230E1, and surge protection wires (m)	Supply cables meeting Rule 230C1, and supply cables of 0 to 750 V meeting Rule 230C2 or 230C3 (m)	Open supply conductors 0 to 750 V, supply cables over 750 V meeting Rule 230C2 or 230C3 (m)	Open supply conductors over 750 V to 22 kV (m)
1. Communication conductors and cables, and messengers	0.60 (2)	0.60	0.60	1.20 (8)	1.80 (5)
2. Supply cables meeting Rule 230C1, and supply cables of 0 to 750 V meeting Rules 230C2 or 230C3	0.60	0.60	0.60	0.60	0.60

3. Open supply conductors, 0 to 750 V; supply cables over 750 V meeting Rules 230C2 or 230C3	1.20 ⑨	0.60	1.20	**0.60**	0.60
4. Open supply conductors					
750 V to 22 kV	**1.80** ⑤⑨	1.20	1.20 ⑨	1.20 ⑨	0.60
over 22 kV to 50 kV	**1.80** ⑨	1.20	**1.80** ⑨	**1.80** ⑨	1.20 ⑨
5. Trolley and electrified railroad contact conductors and associated span and messenger wires	1.20 ③	1.20 ③	1.20 ③	1.20 ③④	2.40
6. Guys ⑦, span wires, neutral conductors meeting Rule 230E1, and surge protection wires	0.60 ②	0.60 ①②	0.60 ②	0.60	1.20

① This clearance may be reduced where both guys are electrically interconnected.

② The clearance of communication conductors and their guy, span, and messenger wires from each other in locations where no other classes of conductors are involved may be reduced by mutual consent of the parties concerned, subject to the approval of the regulatory body having jurisdiction, except for fire-alarm conductors and conductors used in the operation of railroads, or where one set of conductors is for public use and the other used in the operation of supply systems.

③ Trolley and electrified railroad contact conductors of more than 750 V should have at least 6 ft clearance. This clearance should also be provided over lower voltage trolley and electrified railroad contact conductors unless the crossover conductors are beyond reach of a

(continued on next page)

Footnotes for Table 233-1 *(Continued)*

trolley pole leaving the trolley-contact conductor or are suitably protected against damage from trolley poles leaving the trolley-contact conductor.

④ Trolley and electrified railroad feeders are exempt from this clearance requirement for contact conductors if they are of the same nominal voltage and of the same system.

⑤ This clearance may be reduced to 1.20 m where supply conductors of 750 V to 8.7 kV cross a communication line more than 1.80 m horizontally from a communication structure.

⑥ This footnote not used in this edition.

⑦ These clearances may be reduced by not more than 25% to a guy insulator, provided that full clearance is maintained to its metallic end fittings and the guy wires. The clearance to an insulated section of a guy between two insulators may be reduced by not more than 25% provided that full clearance is maintained to the uninsulated portion of the guy.

⑧ This clearance may be reduced to 0.60 m for supply service drops.

⑨ In general, this type of crossing is not recommended.

Table 233-2 Clearance Between Supply Wires, Conductors, and Cables in Rule 233C3b(1) **FT**

(Add 3% for each 1000 ft in excess of 1500 ft above mean sea level.)

Higher voltage circuit		Lower voltage circuit						
Maximum operating voltage phase to phase (kV)	Switching surge factor (per unit)	Maximum operating voltage, phase to phase (kV) 121 (ft)	145 (ft)	169 (ft)	242 (ft)	362 (ft)	550 (ft)	800 (ft)
242	3.3 or less	① 7.0	① 7.0	① 7.0	① 7.1			
362	2.4	① 9.3	① 9.3	① 9.3	① 9.3	9.4		
	2.6	① 9.3	① 9.3	① 9.3	① 9.3	10.3		
	2.8	① 9.3	① 9.3	① 9.3	9.7	11.3		
	3.0	① 9.3	9.4	9.7	10.7	12.3		
550	1.8	① 13.0	① 13.0	① 13.0	① 13.0	① 13.0	13.6	
	2.0	① 13.0	① 13.0	① 13.0	① 13.0	① 13.0	15.3	
	2.2	① 13.0	① 13.0	① 13.0	① 13.0	14.1	17.0	
	2.4	① 13.0	① 13.0	① 13.0	14.0	15.8	18.8	
	2.6	② 13.6	② 14.1	14.5	15.6	17.5	20.7	
800	1.6	① 17.8	① 17.8	① 17.8	① 17.8	① 17.8	18.2	22.5
	1.8	① 17.8	① 17.8	① 17.8	① 17.8	① 17.8	20.9	25.4
	2.0	① 17.8	① 17.8	① 17.8	18.4	20.4	23.7	28.5
	2.2	② 18.4	② 18.9	② 19.4	② 20.8	② 23.1	② 26.7	② 31.5

① Limited by Rule 233C3c.
② Need not be greater than the values specified in Rules 233C1 and 233C2.

Table 233-2 Clearance Between Supply Wires, Conductors, and Cables in Rule 233C3b(1) **M**

(Add 3% for each 300 m in excess of 450 m above mean sea level.)

Higher voltage circuit		Lower voltage circuit						
Maximum operating voltage phase to phase (kV)	Switching surge factor (per unit)	Maximum operating voltage, phase to phase (kV)						
		121 (m)	145 (m)	169 (m)	242 (m)	362 (m)	550 (m)	800 (m)
242	3.3 or less	① 2.13	① 2.13	① 2.13	① 2.16			
362	2.4	① 2.80	① 2.80	① 2.80	① 2.80	2.90		
	2.6	① 2.80	① 2.80	① 2.80	① 2.80	3.1		
	2.8	① 2.80	① 2.80	① 2.80	3.0	3.4		
	3.0	① 2.80	2.90	3.0	3.3	3.7		
550	1.8	① 4.0	① 4.0	① 4.0	① 4.0	① 4.0	4.1	
	2.0	① 4.0	① 4.0	① 4.0	① 4.0	① 4.0	4.7	
	2.2	① 4.0	① 4.0	① 4.0	① 4.0	4.3	5.2	
	2.4	① 4.0	① 4.0	① 4.0	4.3	4.8	5.7	
	2.6	① 4.1	② 4.3	4.4	4.8	5.3	6.3	
800	1.6	① 5.4	① 5.4	① 5.4	① 5.4	① 5.4	5.6	6.9
	1.8	① 5.4	① 5.4	① 5.4	① 5.4	① 5.4	6.4	7.7
	2.0	① 5.4	① 5.4	① 5.4	5.6	6.2	7.0	8.4
	2.2	① 5.6	② 5.8	② 5.9	② 6.3	② 7.0	② 8.1	② 9.4

①Limited by Rule 233C3c.
②Need not be greater than the values specified in Rules 233C1 and 233C2.

Table 233-3 Reference Heights

Reference height	(ft)	(m)
(1) Supply lines	0	0
(2) Communication Lines	2	0.60

234. Clearance of Wires, Conductors, Cables, and Equipment from Buildings, Bridges, Rail Cars, Swimming Pools, and Other Installations

A. Application

1. Vertical and Horizontal Clearances (No Wind Displacement)

   The vertical and horizontal clearances specified in Rules 234B, 234C, 234D, 234E, 234F, and 234I apply under whichever conditions of the following conductor temperature and loading conditions produces the closest approach. Rules 234A1a, 234A1b, and 234A1c apply above and alongside subject installations; Rule 234A1d applies below and alongside subject installations.

   a. 120 °F (50°C), no wind displacement, final sag.

   b. The maximum conductor temperature for which the line is designed to operate, if greater than 120 °F (50 °C), no wind displaceement, final sag.

   c. 32 °F (0 °C), no wind displacement, final sag, with radial thickness of ice, if any, specified in Rule 250B for the applicable loading district.

   d. The minimum conductor temperature for which the line is designed, no wind displacement, initial sag.

   EXCEPTION: Vertical or lateral conductors or cables attached directly to the surface of a supporting structure in accordance with other rules are not subject to the provisions of this rule.

   NOTE: The phase and neutral conductors of a supply line are normally considered separately when determining the sag of each due to temperature rise.

2. Horizontal Clearances (With Wind Displacement)

   Where consideration of horizontal displacement under wind conditions is required, the conductors or cables shall be considered to be displaced from rest toward the installation by a 6 pounds per square foot (290 Pa) wind at final sag at 60 °F (15 °C). This may be reduced to a 4 pounds per square foot (190 Pa) wind in areas sheltered by buildings, terrain, or other obstacles. The displacement of a conductor or cable shall include deflection of suspension insulators and flexible structures.

EXCEPTION: Deflection of flexible structures need not be considered if the highest conductor or shield wire support is less than 60 ft (10 m) above grade.

3. Transition Between Horizontal and Vertical Clearances
The horizontal clearance governs above the level of the roof or top of an installation to the point where the diagonal equals the vertical clearance requirement. Similarly, the horizontal clearance governs above or below projections from buildings, signs, or other installations to the point where the diagonal equals the vertical clearance requirement. From this point the transitional clearance shall equal the vertical clearance as shown in Figs 234-1(a) and 234-1(b). This rule should not be interpreted as restricting the installation of a trolley-contact conductor over the approximate center line of the track it serves.

B. Clearances of Wires, Conductors, and Cables from Other Supporting Structures
Wires, conductors, or cables of one line passing near a lighting support, traffic signal support, or a supporting structure of a second line, without being attached thereto, shall have clearance from any part of such structure not less than the following:

1. A horizontal clearance, without wind, of 5 ft (1.50 m) for voltages up to 50 kilovolts.
When the following conductors and cables are displaced from rest under the wind conditions of Rule 234A2, horizontal clearances from such conductors or cables to other supporting structures shall be not less than those shown below.

Conductor or cable	Horizontal clearance required when displaced by wind (ft)	(m)
Open supply conductor, 0 – 750 V	3.5	1.1
230C2 cable, above 750 V	3.5	1.1
230C3 cable, above 750 V	3.5	1.1
Open supply conductors, 751 V – 22 kV	4.5	1.5

See footnotes 9 and 10 to Table 234-1.

2. A vertical clearance of 4.5 ft (1.40 m) for voltages below 22 kilovolts and a vertical clearance of 5.5 ft (1.70 m) for voltages between 22 and 50 kilovolts.
EXCEPTION 1: For guys, messengers, and neutrals meeting Rule 230E1 and for cables of 300 V or less to ground meeting the requirements of Rule 230C1, 230C2, or 230C3, the vertical clearance may be reduced

to 2 ft (600 mm) and the horizontal clearance may be reduced to 3 ft (900 mm).

NOTE: Clearances of wires, conductors, and cables from adjacent line structure guy wires are given in Rule 233.

EXCEPTION 2: The vertical clearances may be reduced by 2 ft (600 mm) if both of the following conditions are met:

a. The wires, conductors, or cables above and the supporting structure of another line below are operated and maintained by the same utility.
b. Employees do not work above the top of the supporting structure unless:
   (1) The upper circuit is de-energized or temporarily insulated or repositioned, or
   (2) Other equivalent measures are taken.

C. Clearances of Wires, Conductors, Cables, and Rigid Live Parts from Buildings, Signs, Billboards, Chimneys, Radio and Television Antennas, Tanks, and Other Installations Except Bridges

1. Vertical and Horizontal Clearances

a. Clearances

Unguarded or accessible wires, conductors, cables, or rigid live parts may be located adjacent to buildings, signs, billboards, chimneys, radio and television antennas, tanks, and other installations and any projections therefrom. The vertical and horizontal clearances of such rigid and nonrigid parts shall be not less than the values given in Table 234-1 when at rest under the conditions specified in Rule 234A1. These facilities may be installed beside, over or under buildings, building projections and other installations, as illustrated in Figs 234-1(a) and 234-1(b).

b. Horizontal Clearances Under Wind Displacement Conditions

When the following conductors and cables are displaced from rest under the wind conditions of Rule 234A2, horizontal clearances from such conductors or cables to buildings, signs, billboards, chimneys, radio and television antennas, and other installations shall be not less than those shown below.

Conductor or cable	Horizontal clearance required when displaced by wind (ft)	(m)
Open supply conductor, 0–750 V	3.5	1.1
230C2 cable, above 750 V	3.5	1.1
230C3 cable, above 750 V	3.5	1.1
Open supply conductors, 751 V–22 kV	4.5	1.5

See footnotes 9 and 10 to Table 234-1.

2. Guarding of Supply Conductors
   Where the clearances set forth in Table 234-1 cannot be obtained, supply conductors shall be guarded.
   NOTE: Metal-clad supply cables meeting Rule 230C1 are considered to be guarded within the meaning of this rule.
3. Supply Conductors Attached to Buildings or Other Installations
   Where the permanent attachment of supply conductors of any class is necessary for an entrance, such conductors shall meet the following requirements over or along the installation to which the conductors are attached:
   a. Conductors of more than 300 volts to ground shall not be carried along or near the surface of the installation unless they are guarded or made inaccessible.
   b. Clearance of wires from the surface of the installation shall be not less than those required in Table 235-6 (Rule 235E1) for clearance of conductors from supports.
   c. Service-drop conductors shall not be readily accessible and when not in excess of 750 volts they shall have a clearance of not less than the following:
      (1) 8 ft (2.45 m) from the highest point of roofs or balconies over which they pass.
          EXCEPTION 1: Where the voltage between conductors does not exceed 300 volts or where the voltage of cables meeting Rule 230C2 or 230C3 does not exceed 750 volts and the roof or balcony is not readily accessible, the clearance may be not less than 2 ft (600 mm). A roof or balcony is considered readily accessible if the means of access is through a doorway, window, ramp, stairway or permanently mounted ladder. A permanently mounted ladder is not considered a means of access if its bottom rung is 8 ft (2.45 m) or more from the ground or other permanently installed accessible surface.
          EXCEPTION 2: Where not more than 4 ft (1.20 m) of a service drop, measured horizontally, meeting one of the following conditions pass over a roof to terminate at a (through-the-roof) service raceway or approved support, the clearance above the roof may be maintained at not less than 18 in (450 mm).

(a) Conductors of 300 volts or less, or

(b) Cables of 750 volts or less meeting Rules 230C2 or 230C3.

(2) 3 ft (900 mm) in any direction from windows, doors, porches, fire escapes, or similar locations.

EXCEPTION 1: This does not apply to service-drop conductors meeting Rule 230C3 above the top level of a window.

EXCEPTION 2: This does not apply to windows that are not designed to open.

4. Communications Conductors Attached to Buildings or Other Installations

Communications conductors and cables may be attached directly to buildings or other installations.

5. Ladder Space

Where buildings or other installations exceed three stories [or 50 ft (15 m)] in height, overhead lines should be arranged where practical so that a clear space or zone at least 6 ft (1.80 m) wide will be left either adjacent to the building or beginning not over 8 ft (2.45 m) from the building to facilitate the raising of ladders where necessary for fire fighting.

EXCEPTION: This requirement does not apply where it is the unvarying rule of the local fire departments to exclude the use of ladders in alleys or other restricted places which are generally occupied by supply conductors and cables.

D. Clearance of Supply Wires, Conductors, Cables, and Rigid Live Parts from Bridges

1. Vertical and Horizontal Clearances

a. Clearances

Unguarded or accessible wires, conductors, cables, or rigid live parts may be located adjacent to or within a bridge structure. The vertical and horizontal clearances of such rigid and nonrigid parts shall be not less than the values given in Table 234-2 when at rest under the conditions specified in Rule 234A1, as illustrated in Figs 234-1a and 234-1b.

EXCEPTION: This rule does not apply to guys, span wires, effectively grounded surge protection wires, neutrals meeting Rule 230E1, and supply cables meeting Rule 230C1.

b. Horizontal Clearances Under Wind Displacement Conditions

When the following conductors and cables are displaced from rest under the wind conditions of Rule 234A2, horizontal clearances from such conductors or cables to bridges shall be not less than those shown below.

Conductor or cable	Horizontal clearance required when displaced by wind (ft)	(m)
Open supply conductor, 0–750 V	3.5	1.1
230C2 cable, above 750 V	3.5	1.1
230C3 cable, above 750 V	3.5	1.1
Open supply conductors, over 750 V–22 kV	4.5	1.5

See footnotes 8 and 9 to Table 234-2.

2. Guarding Trolley-Contact Conductors Located Under Bridges

a. Where Guarding is Required

Guarding is required where the trolley-contact conductor is located so that a trolley pole leaving the conductor can make simultaneous contact between it and the bridge structure.

b. Nature of Guarding

Guarding shall consist of a substantial inverted trough of nonconducting material located above the contact conductor, or of other suitable means of preventing contact between the trolley support and the bridge structure.

E. Clearance of Wires, Conductors, or Cables Installed Over or Near Swimming Areas

1. Swimming Pools

Where wires, conductors, or cables cross over a swimming pool or the surrounding area within 25 ft (7.6 m) of the edge of the pool, the clearances in any direction shall be not less than those shown in Table 234-3 and illustrated in Fig 234-2. Horizontal clearances shall be met under the wind conditions of Rule 234A2.

EXCEPTION 1: This rule does not apply to a pool fully enclosed by a solid or screened permanent structure.

EXCEPTION 2: This rule does not apply to communication conductors and cables, effectively grounded surge protection wires, neutral conductors meeting Rule 230E1, guys and messengers, supply cables meeting Rule 230C1, supply cables of 0 to 750 V meeting Rules 230C2 or 230C3; when these facilities are 10 ft (3.0 m) or more horizontally from the edge of the pool, diving platform, or diving tower.

2. Beaches and Waterways Restricted to Swimming

Where rescue poles are used by lifeguards at supervised swimming beaches, the vertical and horizontal clearances shall be not less than those shown in Table 234-3. Where rescue poles are not used, the clearances shall be as specified in Rule 232.

3. Waterways Subject to Water Skiing

The vertical clearance shall be the same as that specified in Rule 232.

F. Clearances of Wires, Conductors, Cables, and Rigid Live Parts from Grain Bins

1. Grain Bins Loaded by Permanently Installed Augers, Conveyers or Elevator Systems

All portions of grain bins that are expected to be loaded by the use of a permanently installed auger, conveyer, or elevator system may be considered as a building or other installation under Rule 234C for the purpose of determining appropriate clearances of wires, conductors, cables, and rigid live parts, except that a vertical clearance above the bin of not less than 18 ft (5.5 m) shall be maintained above the level of the highest probe port.

2. Grain Bins Loaded by Portable Augers, Conveyers or Elevators

a. The clearance of wires, conductors, cables, and rigid live parts from grain bins that are expected to be loaded by the use of a portable auger, conveyer, or elevator shall be not less than the values illustrated in Figure 234-3. Horizontal clearances shall be met under the wind conditions of Rule 234A2.

EXCEPTION: Clearances to grain bins of the categories of the following items shall be not less than those required by Rule 234C for clearances from buildings:

(1) Support arms; effectively grounded equipment cases

(2) Insulated communication conductors and cables, messengers, surge protection wires, grounded guys, neutral conductors meeting Rule 230E1, and supply cables meeting Rule 230C1

(3) Unguarded rigid live parts of 0 to 750V, non-insulated communication conductors, supply cables of 0–750 V meeting Rules 230C2 or 230C3, and unguarded cases of equipment at 0 to 750 V.

b. Any side of a grain bin is considered to be a non-loading side if it is so designated, or if it is so closely abutting another structure or obstruction, or so close to a public road or other right-of-way that a portable auger, conveyor, or elevator is not reasonably anticipated to be used over that side or portion to fill the grain bin.

c. Where an agreement excludes the use of portable augers, conveyors, or elevators from a designated portion of a grain bin, such portion is considered to be a non-loading side.

G. Additional Clearances for Voltages Exceeding 22 kV for Wires, Conductors, Cables, and Unguarded Rigid Live Parts of Equipment

Greater clearances than specified in Rules 234B, 234C, 234D, 234E, 234F, and 234J shall be provided where required below.

1. For voltages between 22 and 470 kilovolts, the clearance specified in Rules 234B, 234C, 234D, 234E, 234F, and 234J shall be increased at the rate of 0.4 inch (10 mm) per kilovolt in excess of 22 kilovolts. For voltages exceeding 470 kilovolts, the clearance shall be determined by the method given in Rule 234H. All clearances for lines over 50 kilovolts shall be based on the maximum operating voltage.

   EXCEPTION 1: Where a clearance value is given for the 22 kV to 50 kV range, the voltage adder of Rule 234G1 applies to the voltage in excess of 50 kV.

   EXCEPTION 2: For voltages exceeding 98 kilovolts alternating current to ground or 139 kilovolts direct current to ground, clearances less than those required above are permitted for systems with known maximum switching surge factor. (See Rule 234H.)

2. For voltages exceeding 50 kV, the additional clearance specified in Rule 234G1 shall be increased 3% for each 1000 ft (300 m) in excess of 3300 ft (1000 m) above mean sea level.

3. For voltages exceeding 98 kilovolts alternating current to ground, or 139 kilovolts direct current to ground, either

the clearances shall be increased or the electric field, or the effects thereof, shall be reduced by other means, as required, to limit the current due to electrostatic effects to 5.0 milliamperes, rms, if an ungrounded metal fence, building, sign, billboard, chimney, radio or television antenna, tank or other installation, or any ungrounded metal attachments thereto were short-circuited to ground. For this determination, the conductor shall be at a final unloaded sag at 120 °F (50 °C).

H. Alternate Clearances for Voltages Exceeding 98 Kilovolts Alternating Current to Ground or 139 Kilovolts Direct Current to Ground

The clearances specified in Rules 234B, 234C, 234D, 234E, 234F, 234G, and 234J may be reduced for circuits with known switching surge factors, but shall be not less than the alternate clearance, which is computed by adding the reference distance from Rule 234H2 to the electrical component of clearance from Rule 234H3.

1. Sag Conditions of Line Conductors

   The vertical, horizontal, and diagonal clearances shall be maintained under the conductor temperature and loading conditions given in Rule 234A.

2. Reference Distances

   The reference distance shall be selected from Table 234-5.

3. Electrical Component of Clearance

   a. The electrical component (D) shall be computed using the following equations. Selected values of D are listed in Table 234-4.

$$D = 3.28 \left[\frac{V \cdot (PU) \cdot a}{500\,K}\right]^{1.667} bc \quad \text{(feet)}$$

$$D = 1.00 \left[\frac{V \cdot (PU) \cdot a}{500\,K}\right]^{1.667} bc \quad \text{(m)}$$

where

V = maximum alternating current crest operating voltage to ground or maximum direct current operating voltage to ground in kilovolts;

PU = maximum switching surge factor expressed in per-unit peak voltage to ground and defined as a switching surge level for circuit breakers corresponding to 98% probability that the maximum switching surge generated per breaker operation does not exceed this surge level, or the maximum anticipated switching surge level generated by other means, whichever is greater;

a = 1.15, the allowance for three standard deviations;

b = 1.03, the allowance for nonstandard atmospheric conditions;

c = the margin of safety:

1.2 for vertical clearances

1.0 for horizontal clearances;

K = 1.15, the configuration factor for conductor-to-plane gap.

b. The value of *D* shall be increased 3% for each 1000 ft (300 m) in excess of 1500 ft (450 m) above mean sea level.

4. Limit

The alternate clearance shall be not less than the clearance of Rule 234B, Table 234-1, Table 234-2, or Table 234-3, as applicable, computed for 98 kilovolts alternating current rms to ground by Rule 234G1.

I. Clearance of Wires, Conductors, and Cables to Rail Cars

Where overhead wires, conductors, or cables run along railroad tracks, the clearance in any direction shall be not less than that shown in Fig 234-4. The values of *V* and *H* are as defined below:

V = vertical clearance from the wire, conductor, or cable above the top of the rail as specified in Rule 232 minus 22 ft (6.7 m), the assumed height of the rail car

H = horizontal clearance from the wire, conductor, or cable to the nearest rail, which is equal to the required vertical clearance above the rail minus 15 ft (4.6 m) as computed by the lesser of the following:

1. Rules 232B1 and 232C1
2. Rule 232D.

These clearances are computed for railroads handling standard rail cars as common carriers in interchange service with other railroads. Where wires, conductors, or cables run along mine, logging, and similar railways which handle only cars smaller than standard freight cars, the value of H may be reduced by one-half the difference between the width of a standard rail car [10 ft, 8 in (3.3 m)] and the width of the narrower car.

J. Clearance of Equipment Mounted on Supporting Structures

1. Clearance to Unguarded Rigid Live Parts of Equipment
   The horizontal and vertical clearances of unguarded rigid live parts such as potheads, transformer bushings, surge arresters, and short lengths of supply conductors connected thereto, which are not subject to variation in sag, shall be not less than those required for open supply conductors by Rules 234C or 234D, as applicable.
2. Clearance to Equipment Cases
   Equipment shall be mounted so that clearances are not less than that given by Rules 234J2a, 234J2b, and 234J2c.
   a. Effectively grounded equipment cases may be located on or adjacent to buildings, bridges, or other structures provided that clearances for unguarded rigid live parts of such equipment, as specified in Rule 234J1, are maintained.
   b. Equipment cases which are not effectively grounded shall be located so that the clearances for open supply conductors of Rules 234C or 234D, as applicable, are maintained.
   c. Equipment cases shall be located so as not to serve as a means of approach to unguarded rigid live parts by unqualified persons.

NOTE: Rule 234J is not subject to the loading conditions of Rule 234A.

Fig 234-1(a)
Clearance Diagram for Building

Fig 234-1(b)
Clearance Diagram for Other Structures

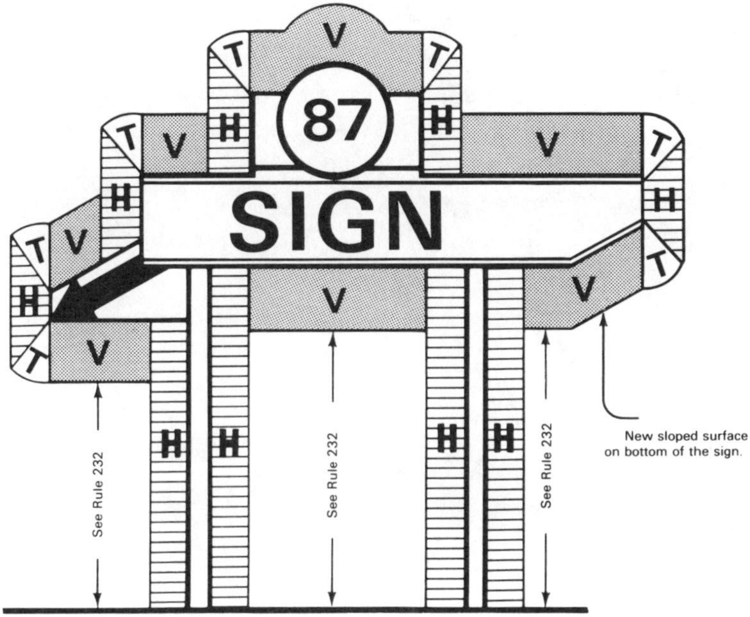

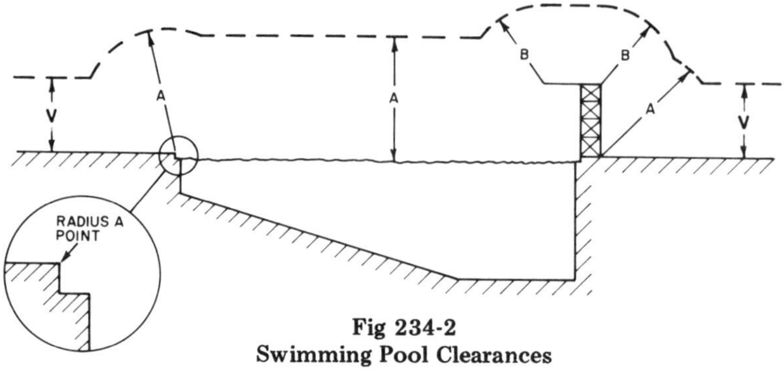

Fig 234-2
Swimming Pool Clearances

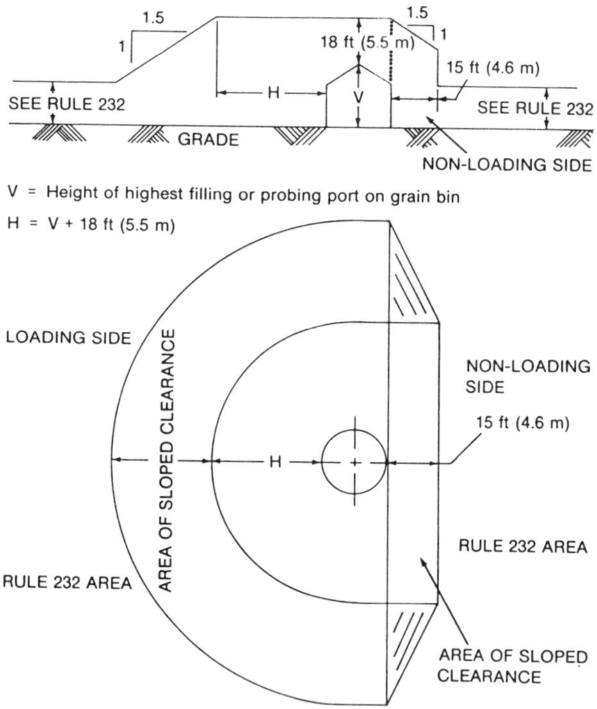

Fig 234-3
Clearance Envelope for Grain Bins Filled by Portable Augers, Conveyors, or Elevators

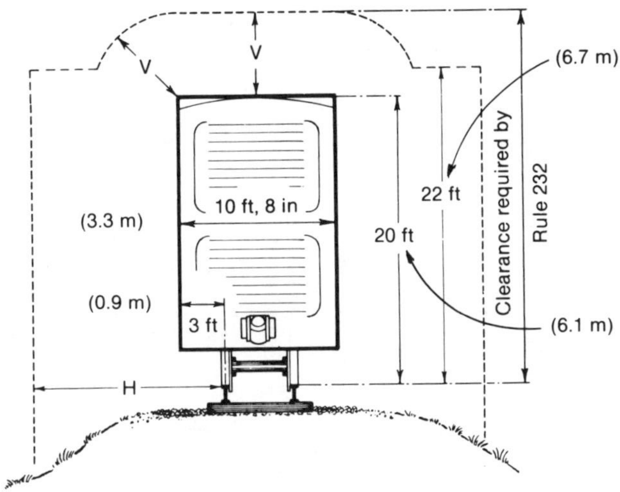

Fig 234-4
Rail Car Clearances

Table 234-1 Clearance of Wires, Conductors, Cables, and Unguarded Rigid Live Parts Adjacent but Not Attached to Buildings and Other Installations Except Bridges **FT**

(Voltages are phase to ground for effectively grounded circuits and those other circuits where all ground faults are cleared by promptly de-energizing the faulted section, both initially and following subsequent breaker operations. See the definitions section for voltages of other systems.)

Clearance of	Insulated communication conductors and cables; messengers; surge protection wires; grounded guys; neutral conductors meeting Rule 230E1; supply cables meeting Rule 230C1 (ft)	Unguarded rigid live parts, 0 to 750 V; non-insulated communication conductors; supply cables of 0 to 750 V meeting Rules 230C2 or 230C3 (ft)	Supply cables over 750 V meeting Rules 230C2 or 230C3; open supply conductors, 0 to 750 V (ft)	Open supply conductors, over 750 V to 22 kV (ft)	Unguarded rigid live parts, over 750 V to 22 kV (ft)
1. Buildings					
a. Horizontal					
(1) To walls, projections and guarded windows	4.5 ⑦	5.0	5.5 ①②⑨	7.5 ①②⑩⑪	7.0
(2) To unguarded windows ⑧	4.5	5.0	5.5 ①②⑨	7.5 ⑩⑪	7.0
(3) To balconies and areas accessible to pedestrians ③	4.5	5.0	5.5 ⑨	7.5 ⑩⑪	7.0
b. Vertical					
(1) Over or under roofs or projections not accessible to pedestrians ③	3.0	3.5	10.5	12.5	12.0

(2) Over or under balconies and roofs accessible to pedestrians ③	10.5	11.0	11.5	13.5	13.0
(3) Over roofs accessible to vehicles but not subject to truck traffic ⑥	10.5	11.0	11.5	13.5	13.0
(4) Over roofs accessible to truck traffic ⑥	15.5	16.0	16.5	18.5	18.0
2. Signs, chimneys, billboards, radio and television antennas, tanks, and other installations not classified as buildings or bridges					
a. Horizontal ④	3.0	3.5	5.5 ①②⑨	7.5 ①②⑩⑪	7.0
b. Vertical over or under ④	3.0	3.5	6.0 ①	8.0	7.5

① Where building, sign, chimney, antenna, tank, or other installation does not require maintenance such as painting, washing, changing of sign letters, or other operations which would require persons to work or pass between supply conductors and structure, the clearance may be reduced by 2 ft.

② Where available space will not permit this value, the clearance may be reduced by 2 ft provided the conductors, including splices and taps, have covering which provides sufficient dielectric to prevent a short circuit in case of momentary contact between the conductors and a grounded surface.

③ A roof, balcony, or area is considered accessible to pedestrians if the means of access is through a doorway, ramp, window, stairway, or permanently mounted ladder. A permanently mounted ladder is not considered a means of access if its bottom rung is 8 ft or more from the ground or other permanently installed accessible surface.

④ **The required clearances shall be to the closest approach of motorized signs or moving portions of installations covered by Rule 234C.**

⑤ This footnote not used in this edition.

⑥ For the purpose of this rule, trucks are defined as any vehicle exceeding 8 ft in height.

⑦ This clearance may be reduced to 3 in for the grounded portions of guys.

⑧ Windows not designed to open may have the clearances permitted for walls and projections.

⑨ This clearance shall be not less than 3.5 ft with the conductor or cable displaced by wind; see Rule 234C1b.

⑩ This clearance shall be not less than 4.5 ft with the conductor displaced by wind; see Rule 234C1b.

⑪ Where available space will not permit this value, the clearance may be reduced to 7.0 ft for conductors limited to 8.7 kV to ground.

Table 234-1 Clearance of Wires, Conductors, Cables, and Unguarded Rigid Live Parts Adjacent but Not Attached to Buildings and Other Installations Except Bridges M

(Voltages are phase to ground for effectively grounded circuits and those other circuits where all ground faults are cleared by promptly de-energizing the faulted section, both initially and following subsequent breaker operations. See the definitions section for voltages of other systems.)

Clearance of	Insulated communication conductors and cables; messengers; surge protection wires; grounded guys; neutral conductors meeting Rule 230E1; supply cables meeting Rule 230C1 (m)	Unguarded rigid live parts, 0 to 750 V; non-insulated communication conductors; supply cables of 0 to 750 V meeting Rules 230C2 or 230C3 (m)	Supply cables over 750 V meeting Rules 230C2 or 230C3; open supply conductors, 0 to 750 V (m)	Open supply conductors, over 750 V to 22 kV (m)	Unguarded rigid live parts, over 750 V to 22 kV (m)
1. Buildings					
a. Horizontal					
(1) To walls, projections and guarded windows	1.40 ⑦	1.50	1.70 ①②⑨	2.30 ①②⑩⑪	2.00
(2) To unguarded windows ⑧	1.40	1.50	1.70 ①②⑨	2.30 ⑩⑪	2.00
(3) To balconies and areas accessible to pedestrians ③	1.40	1.50	1.70 ⑨	2.30 ⑩⑪	2.00
b. Vertical					
(1) Over or under roofs or projections not accessible to pedestrians ③	0.90	1.07	3.2	3.8	3.6

(2) Over or under balconies and roofs accessible to pedestrians ③	3.2	3.4	3.5	4.1	4.0
(3) Over roofs accessible to vehicles but not subject to truck traffic ⑥	3.2	3.4	3.5	4.1	4.0
(4) Over roofs accessible to truck traffic ⑥	4.7	4.9	5.0	5.6	5.5
2. Signs, chimneys, billboards, radio and television antennas, tanks, and other installations not classified as buildings or bridges					
a. Horizontal ④	0.90	1.07	1.70 ①②⑨	2.30 ①②⑩⑪	2.00
b. Vertical over or under ④	0.90	1.07	1.80 ①	2.45	2.30

① Where building, sign, chimney, antenna, tank, or other installation does not require maintenance such as painting, washing, changing of sign letters, or other operations which would require persons to work or pass between supply conductors and structure, the clearance may be reduced by 0.60 m.

② Where available space will not permit this value, the clearance may be reduced by 0.60 m provided the conductors, including splices and taps, have covering which provides sufficient dielectric to prevent a short circuit in case of momentary contact between the conductors and a grounded surface.

③ A roof, balcony, or area is considered accessible to pedestrians if the means of access is through a doorway, ramp, window, stairway, or permanently mounted ladder. A permanently mounted ladder is not considered a means of access if its bottom rung is 2.45 m or more from the ground or other permanently installed accessible surface.

④ The required clearances shall be to the closest approach of motorized signs or moving portions of installations covered by Rule 234C.

⑤ This footnote not used in this edition.

⑥ For the purpose of this rule, trucks are defined as any vehicle exceeding 2.45 m in height.

⑦ This clearance may be reduced to 75 mm for the grounded portions of guys.

⑧ Windows not designed to open may have the clearances permitted for walls and projections.

⑨ This clearance shall be not less than 1.07 m with the conductor or cable displaced by wind; see Rule 234C1b.

⑩ This clearance shall be not less than 1.40 m with the conductor displaced by wind; see Rule 234C1b.

⑪ Where available space will not permit this value, the clearance may be reduced to 2.00 m for conductors limited to 8.7 kV to ground.

Table 234-2 Clearance of Supply Conductors, Cables, and Unguarded Rigid Live Parts from Bridges **FT**

(Voltages are phase to ground for effectively grounded circuits and those other circuits where all ground faults are cleared by promptly de-energizing the faulted section, both initially and following subsequent breaker operations. See the definitions section for voltages of other systems.)

	Unguarded rigid live parts, 0 to 750 V; non-insulated communication conductors; supply cables of 0 to 750 V meeting Rules 230C2 or 230C3 (ft)	Supply cables over 750 V meeting Rules 230C2 or 230C3; open supply conductors, 0 to 750 V (ft)	Open supply conductors, over 750 V to 22 kV (ft)	Unguarded rigid live parts, over 750 V to 22 kV (ft)
1. Clearance over bridges (1)				
a. Attached (3)	3.0	3.5	5.5	5.0
b. Not attached	10.0	10.5	12.5	12.0
2. Clearance beside, under, or within bridge structure (6)				
a. Readily accessible portions of any bridge including wing, walls, and bridge attachments (1)				
(1) Attached (3)	3.0	3.5	5.5	5.0
(2) Not attached	5.0	5.5 (8)	7.5 (9)	7.0
b. Ordinarily inaccessible portions of bridges (other than brick, concrete, or masonry) and from abutments (2)				
(1) Attached (3)(5)	3.0	3.5	5.5	5.0
(2) Not attached (4)(5)	4.0	4.5 (8)	6.5 (9)	6.0

Footnotes for Table 234-2

① Where over traveled ways on or near bridges, the clearances of Rule 232 apply also.

② Bridge seats of steel bridges carried on masonry, brick, or concrete abutments which require frequent access for inspection shall be considered as readily accessible portions.

③ Clearance from supply conductors to supporting arms and brackets attached to bridges shall be the same as specified in Table 235-6 (Rule 235E1) if the supporting arms and brackets are owned, operated, or maintained by the same utility.

④ This footnote not used in this edition.

⑤ Where conductors passing under bridges are adequately guarded against contact by unauthorized persons and can be de-energized for maintenance of the bridge, clearances of the conductors from the bridge, at any point, may have the clearances specified in Table 235-6 for clearance from surfaces of support arms plus one-half the final unloaded sag of the conductor at that point.

⑥ Where the bridge has moving parts, such as a lift bridge, the required clearances shall be maintained throughout the full range of movement of the bridge or any attachment thereto.

⑦ Where permitted by the bridge owner, supply cables may be run in rigid conduit attached directly to the bridge. Refer to Part 3 for installation rules.

⑧ This clearance shall be not less than 3.5 ft with the conductor or cable displaced by wind; see Rule 234D1b.

⑨ This clearance shall be not less than 4.5 ft with the conductor displaced by wind; see Rule 234D1b.

Table 234-2 Clearance of Supply Conductors, Cables, and Unguarded Rigid Live Parts from Bridges **M**

(Voltages are phase to ground for effectively grounded circuits and those other circuits where all ground faults are cleared by promptly de-energizing the faulted section, both initially and following subsequent breaker operations. See the definitions section for voltages of other systems.)

	Unguarded rigid live parts, 0 to 750 V; non-insulated communication conductors; supply cables of 0 to 750 V meeting Rules 230C2 or 230C3 (m)	Supply cables over 750 V, meeting Rules 230C2 or 230C3; open supply conductors, 0 to 750 V (m)	Open supply conductors, over 750 V to 22 kV (m)	Unguarded rigid live parts, over 750 V to 22 kV (m)
1. Clearance over bridges ①				
a. Attached ③	0.90	1.07	1.70	1.50
b. Not attached	3.0	3.2	3.8	3.6
2. Clearance beside, under, or within bridge structure ⑥				
a. Readily accessible portions of any bridge including wing, walls, and bridge attachments ①				
(1) Attached ③	0.90	1.07	1.70	1.50
(2) Not attached	1.50	1.70 ⑧	2.30 ⑨	2.00
b. Ordinarily inaccessible portions of bridges (other than brick, concrete, or masonry) and from abutments ②				
(1) Attached ③⑤	0.90	1.07	1.70	1.50
(2) Not attached ④⑤	1.20	1.40 ⑧	2.00 ⑨	1.80

Footnotes for Table 234-2

① Where over traveled ways on or near bridges, the clearances of Rule 232 apply also.

② Bridge seats of steel bridges carried on masonry, brick, or concrete abutments which require frequent access for inspection shall be considered as readily accessible portions.

③ Clearance from supply conductors to supporting arms and brackets attached to bridges shall be the same as specified in Table 235-6 (Rule 235E1) if the supporting arms and brackets are owned, operated, or maintained by the same utility.

④ This footnote not used in this edition.

⑤ Where conductors passing under bridges are adequately guarded against contact by unauthorized persons and can be de-energized for maintenance of the bridge, clearances of the conductors from the bridge, at any point, may have the clearances specified in Table 235-6 for clearance from surfaces of support arms plus one-half the final unloaded sag of the conductor at that point.

⑥ Where the bridge has moving parts, such as a lift bridge, the required clearances shall be maintained throughout the full range of movement of the bridge or any attachment thereto.

⑦ Where permitted by the bridge owner, supply cables may be run in rigid conduit attached directly to the bridge. Refer to Part 3 for installation rules.

⑧ This clearance shall be not less than 1.07 m with the conductor or cable displaced by wind; see Rule 234D1b.

⑨ This clearance shall be not less than 1.40 m with the conductor displaced by wind; see Rule 234D1b.

Table 234-3 Clearance of Wires, Conductors, and Cables Passing Over or Near Swimming Areas

(Voltages are phase to ground for effectively grounded circuits and those other circuits where all ground faults are cleared by promptly de-energizing the faulted section, both initially and following subsequent breaker operations. See the definitions section for voltages of other systems.)

	Insulated communication conductors and cables; messengers; surge protection wires; grounded guys; neutral conductors meeting Rule 230E1; supply cables meeting Rule 230C1 (ft)	Unguarded rigid live parts, 0 to 750 V; non-insulated communication conductors; supply cables of 0 to 750 V meeting Rules 230C2 or 230C3 (ft)	Supply cables over 750 V meeting Rules 230C2 or 230C3; open supply conductors, 0 to 750 V (ft)	Open supply conductors, over 750 V to 22 kV (ft)
A. Clearance in any direction from the water level, edge of pool, base of diving platform, or anchored raft	22.0	22.5	23.0	25.0
B. Clearance in any direction to the diving platform or tower	14.0	14.5	15.0	17.0
V. Vertical clearance over adjacent land	Clearance shall be as required by Rule 232.			

NOTE: A, B, and V are shown in Fig 234-2.

M

Table 234-3 Clearance of Wires, Conductors, and Cables Passing Over or Near Swimming Areas

(Voltages are phase to ground for effectively grounded circuits and those other circuits where all ground faults are cleared by promptly de-energizing the faulted section, both initially and following subsequent breaker operations. See the definitions section for voltages of other systems.)

	Insulated communication conductors and cables; messengers; surge protection wires; grounded guys; neutral conductors meeting Rule 230E1; supply cables meeting Rule 230C1 (m)	Unguarded rigid live parts, 0 to 750 V; non-insulated communication conductors; supply cables of 0 to 750 V meeting Rules 230C2 or 230C3 (m)	Supply cables over 750 V meeting Rules 230C2 or 230C3; open supply conductors, 0 to 750 V (m)	Open supply conductors, over 750 V to 22 kV (m)
A. Clearance in any direction from the water level, edge of pool, base of diving platform, or anchored raft	6.7	6.9	7.0	7.6
B. Clearance in any direction to the diving platform or tower	4.3	4.4	4.6	5.2
V. Vertical clearance over adjacent land	Clearance shall be as required by Rule 232.			

NOTE: A, B, and V are shown in Fig 234-2.

Table 234-4 Electrical Component of Clearance of Buildings, Bridges, and Other Installations in Rule 234H3a

(Add 3% for each 1000 ft (300 m) in excess of 1500 ft (450 m) above mean sea level.)

Maximum operating voltage phase to phase (kV)	Switching surge factor (per unit)	Switching surge (kV)	Electrical component of clearances V (ft)	V (m)	H (ft)	H (m)
242	2.0	395	2.7	0.82	2.3	0.70
	2.2	435	3.2	0.98	2.7	0.82
	2.4	474	3.7	1.13	3.1	0.94
	2.6	514	4.2	1.28	3.5	1.07
	2.8	553	4.8	1.40	4.0	1.20
	3.0	593	5.4	1.65	4.5	1.40
362	1.8	532	4.5	1.40	3.7	1.13
	2.0	591	5.4	1.65	4.5	1.40
	2.2	650	6.3	1.90	5.2	1.60
	2.4	709	7.3	2.20	6.1	1.85
	2.6	768	8.3	2.50	6.9	2.10
	2.8	828	9.4	2.90	7.8	2.40
	3.0	887	10.6	3.2	8.8	2.70
550	1.6	719	7.5	2.30	6.2	1.90
	1.8	808	9.1	2.80	7.6	2.30
	2.0	898	10.8	3.3	9.0	2.70
	2.2	988	12.7	3.9	10.6	3.2
	2.4	1079	14.6	4.4	12.2	3.7
	2.6	1168	16.7	5.1	13.9	4.2
800	1.6	1045	13.9	4.2	11.6	3.5
	1.8	1176	16.9	5.2	14.1	4.3
	2.0	1306	20.1	6.1	16.7	5.1
	2.2	1437	23.6	7.2	19.7	6.0
	2.4	1568	27.3	8.3	22.7	6.9

Table 234-5 Reference Distances

	Reference distance	Vertical (ft)	Vertical (m)	Horizontal (ft)	Horizontal (m)
a.	Buildings	9	2.70	3	0.90
b.	Signs, chimneys, radio and television antennas, tanks, and other installations not classified as bridges or buildings	9	2.70	3	0.90
c.	Superstructure of bridges ①, ②	9	2.70	3	0.90
d.	Supporting structures of another line	6	1.80	5	1.50
e.	Dimension *A* of Figure 234-2	18	5.5	—	—
f.	Dimension *B* of Figure 234-2	14	4.3	14	4.3

① Where over traveled ways on or near bridges, the clearances of Rule 232 apply also.

② Where the bridge has moving parts, such as a lift bridge, the required clearances shall be maintained throughout the full range of movement of the bridge or any attachment thereto.

235. Clearance for Wires, Conductors, or Cables Carried on the Same Supporting Structure

A. Application of Rule

1. Multiconductor Wires or Cables

   Cables, and duplex, triple, or paired conductors supported on insulators or messengers, meeting Rules 230C or 230D, whether single or grouped, are for the purposes of this rule considered single conductors even though they may contain individual conductors not of the same phase or polarity.

2. Conductors Supported by Messengers or Span Wires

   Clearances between individual wires, conductors, or cables supported by the same messenger, or between any group and its supporting messenger, or between a trolley feeder, supply conductor, or communication conductor, and their respective supporting span wires, are not subject to the provisions of this rule.

3. Line Conductors of Different Circuits

   a. Unless otherwise stated, the voltage between line conductors of different circuits shall be the greater of the following:

      (1) The phasor difference between the conductors involved

      NOTE: A phasor relationship of 180 degrees is considered appropriate where the actual phasor relationship is unknown.

      (2) The phase-to-ground voltage of the higher voltage circuit.

   b. When the circuits have the same nominal voltage, either circuit may be considered to be the higher voltage circuit.

B. Horizontal Clearance Between Line Conductors

1. Fixed Supports

   Line conductors attached to fixed supports shall have horizontal clearances from each other not less than the larger value required by either Rule 235B1a or Rule 235B1b for the situation concerned. Voltage is between the two conductors for which the clearance is being determined except for railway feeders, which are to ground.

EXCEPTION 1: The pin spacing at buckarm construction may be reduced as specified in Rule 236F to provide climbing space.
EXCEPTION 2: Grades D and N need meet only the requirements of Rule 235B1a.
EXCEPTION 3: These clearances do not apply to cables meeting Rule 230C or covered conductors of the same circuit meeting Rule 230D.
EXCEPTION 4: For voltages to ground exceeding 98 kilovolts alternating current or 139 kilovolts direct current, clearances less than those required by a and b below are permitted for systems with known maximum switching surge factors. (See Rule 235B3.)

a. Horizontal Clearance Between Line Conductors of the Same or Different Circuits

Clearances shall be not less than those given in Table 235-1.

b. Clearance According to Sags

The clearance at the supports of conductors of the same or different circuits of grade B or C shall in no case be less than the values given by the following formulas, at a conductor temperature of 60 °F (15 °C), at final unloaded sag, no wind. The requirements of Rule 235B1a apply if they give a greater separation than this rule.

EXCEPTION: No requirement is specified for clearance between conductors of the same circuit when rated above 50 kilovolts.

In the following, S is the apparent sag in inches of the conductor having the greater sag, and the clearance is in inches.

(1) For line conductors smaller than AWG No 2: clearance = 0.3 in per kilovolt + $7\sqrt{(S/3)-8}$. (Table 235-2 shows selected values up to 46 kV.)

(2) For line conductors of AWG No 2 or larger: clearance = 0.3 in per kilovolt + $8\sqrt{S/12}$. (Table 235-3 shows selected values up to 46 kV.)

(3) For voltages exceeding 814 kilovolts, the clearance shall be determined by the alternate method given by Rule 235B3.

In the following, S is the apparent sag in millimeters of the conductor having the greater sag, and the clearance is in millimeters.

(1) For line conductors smaller than AWG No 2: clearance = 7.6 mm per kilovolt + 7 $\sqrt{8.5S - 5080}$. (Table 235-2 shows selected values up to 46 kV.)

(2) For line conductors of AWG No 2 or larger: clearance = 7.6 mm per kilovolt + 8 $\sqrt{2.12S}$. (Table 235-3 shows selected values up to 46 kV.)

(3) For voltages exceeding 814 kilovolts, the clearance shall be determined by the alternate method given by Rule 235B3.

(4) The clearance for voltages exceeding 50 kilovolts specified in Rule 235B1b(1) and (2) shall be increased 3% for each 1000 ft (300 m) in excess of 3300 ft (1000 m) above mean sea level. All clearances for lines over 50 kilovolts shall be based on the maximum operating voltage.

2. Suspension Insulators

Where suspension insulators are used and are not restrained from movement, the clearance between conductors shall be increased so that one string of insulators may swing transversely throughout a range of insulator swing up to its maximum design swing angle without reducing the values given in Rule 235B1. The maximum design swing angle shall be based on a 6 pounds per sq foot (290 Pa) wind on the conductor at final sag at 60 °F (15 °C). This may be reduced to a 4 pounds per sq foot (190 Pa) wind in areas sheltered by buildings, terrains, or other obstacles. The displacement of the wires, conductors, and cables shall include deflection of flexible structures and fittings, where such deflection would reduce the horizontal clearance between two wires, conductors, or cables.

3. Alternate Clearances for Different Circuits Where One or Both Circuits Exceed 98 Kilovolts Alternating Current to Ground or 139 Kilovolts Direct Current to Ground

The clearances specified in Rules 235B1 and 235B2 may be reduced for circuits with known switching surge factors but shall not be less than the clearances derived from the following computations. For these computations, communication conductors and cables, guys, messengers, neutral conductors meeting Rule 230E1, and supply cables, meeting Rule 230C1 shall be considered line conductors at zero voltage.

a. Clearance

(1) The alternate clearance shall be maintained under the expected loading conditions and shall be not less than the electrical clearance between conductors of different circuits computed from the following equation. For convenience, clearances for typical system voltages are shown in Table 235-4.

$$D = 3.28 \left[\frac{V_{L-L} \cdot (PU) \cdot a}{500\,K}\right]^{1.667} b \text{ (ft)}$$

$$D = 1.00 \left[\frac{V_{L-L} \cdot (PU) \cdot a}{500\,K}\right]^{1.667} b \text{ (m)}$$

where

V_{L-L} = maximum alternating current crest operating voltage in kilovolts between phases of different circuits or maximum direct current operating voltage between poles of different circuits. If the phases are of the same phase and voltage magnitude one phase conductor shall be considered grounded;

PU = maximum switching surge factor expressed in per-unit peak operating voltage between phases of different circuits and defined as a switching surge level between phases for circuit breakers corresponding to 98% probability that the maximum switching surge generated per breaker operation does not exceed this surge level, or the maximum anticipated switching surge level generated by other means, whichever is greater;

a = 1.15, the allowance for three standard deviations;

b = 1.03, the allowance for nonstandard atmospheric conditions;

K = 1.4, the configuration factor for a conductor-to-conductor gap.

(2) The value of D shall be increased 3% for each 1000 ft (300 m) in excess of 1500 ft (450 m) above mean sea level.

b. Limit

The clearance derived from Rule 235B3a shall not be less than the basic clearances given in Table 235-1 computed for 169 kilovolts alternating current.

C. Vertical Clearance Between Line Conductors

All line wires, conductors, and cables located at different levels on the same supporting structure shall have vertical clearances not less than the following.

1. Basic Clearance for Conductors of Same or Different Circuits

The clearances given in Table 235-5 shall apply to line wires, conductors, or cables of 0 to 50 kV attached to supports. No value is specified for clearances between conductors of the same circuit exceeding 50 kV.

EXCEPTION 1: Line wires, conductors, or cables on vertical racks or separate brackets placed vertically and meeting the requirements of Rule 235G may have spacings as specified in that rule.

EXCEPTION 2: Where communication service drops cross under supply conductors on a common crossing structure, the clearance between the communication conductor and an effectively grounded supply conductor may be reduced to 4 in (100 mm) provided the clearance between the communication conductor and supply conductors not effectively grounded meets the requirements of Rule 235C as appropriate.

EXCEPTION 3: Supply service drops of 0–750 V running above and parallel to communication service drops may have a spacing of not less than 12 in (300 mm) at any point in the span including the point of and at their attachment to the building provided the non-grounded conductors are insulated and that the clearance as otherwise required by this rule is maintained between the two service drops at the pole.

EXCEPTION 4: This rule does not apply to conductors of the same circuit meeting Rule 230D.

2. Additional Clearances

Greater clearances than those required for 50 kV in Table 235-5 (Rule 235C1) shall be provided under the following conditions. The increases are cumulative where more than one is applicable.

a. Voltages Exceeding 50 Kilovolts

(1) For voltages between 50 and 814 kilovolts, the clearance between conductors of different circuits shall be increased 0.4 in (10 mm) per kilovolt in excess of 50 kV.

EXCEPTION: For voltages to ground exceeding 98kV alternating current or 139 kV direct current, clearances less than those required above are permitted for systems with known switching surge factors. (See Rule 235C3.)

(2) The increase in clearance for voltages in excess of 50 kV specified in Rule 235C2a(1) shall be increased 3% for each 1000 ft (300 m) in excess of 3300 ft (1000 m) above mean sea level.

(3) All clearances for lines over 50 kV shall be based on the maximum operating voltage.

(4) No value is specified for clearances between conductors of the same circuit.

b. Conductors of Different Sags on Same Support

(1) Line conductors, supported at different levels on the same structures shall have vertical clearances at the supporting structures so adjusted that the clearance at any point in the span shall be not less than any of the following with the upper conductor at final unloaded sag at the maximum temperature for which the conductor is designed to operate, or at final sag, with radial thickness of ice, if any, specified in Rule 250B for the loading district concerned, whichever produces the greater sag, and the lower conductor at final unloaded sag under the same ambient conditions and without electrical loading.

(a) For voltages less than 50 kilovolts between conductors, 75% of that required at the supports by Table 235-5.

EXCEPTION: Neutral conductors meeting Rule 230E1 and supply cables meeting Rule 230C1 running above and parallel to communication cables where the supply neutral or messenger is bonded to the communication messenger, may have a clearance of 12 in (300 mm) at any point in the span provided that a clearance of 30 in

(0.75 m) is maintained between the supply conductors and cables and the communication cables at the supporting poles.

(b) For voltages more than 50 kilovolts between conductors, the value specified in Rule 235C2b(1)(a) increased in accordance with Rule 235C2a.

(2) Sags should be readjusted when necessary to accomplish the foregoing, but not reduced sufficiently to conflict with the requirements of Rule 261H2. In cases where conductors of different sizes are strung to the same sag for the sake of appearance or to maintain unreduced clearance throughout storms, the chosen sag should be such as will keep the smallest conductor involved in compliance with the sag requirements of Rule 261H2.

(3) For span lengths in excess of 150 ft (45 m), vertical clearance at the structure between open supply conductors and communication cables or conductors shall be adjusted so that under conditions of conductor temperature of 60 °F (15 °C), no wind displacement and final unloaded sag, no open supply conductor of over 750 volts but less than 50 kilovolts shall be lower in the span than a straight line joining the points of support of the highest communications cable or conductor.

EXCEPTION: Effectively grounded supply conductors associated with systems of 50 kilovolts or less need meet only the provisions of Rule 235C2b(1).

3. Alternate Clearances for Different Circuits Where One or Both Exceed 98 Kilovolts, Alternating Current, or 139 Kilovolts Direct Current to Ground

The clearances specified in Rules 235C1 and 235C2 may be reduced for circuits with known switching surge factors, but shall not be less than the crossing clearances required by Rule 233C3.

D. Diagonal Clearance Between Line Wires, Conductors, and Cables Located at Different Levels on the Same Supporting Structure

No wire, conductor, or cable may be closer to any other wire, conductor, or cable than defined by the dashed line in Fig 235-1, where V and H are determined in accordance with other parts of Rule 235.

E. Clearances in Any Direction from Line Conductors to Supports, and to Vertical or Lateral Conductors, Span or Guy Wires Attached to the Same Support

1. Fixed Supports

   Clearances shall be not less than those given in Table 235-6.

   EXCEPTION: For voltages exceeding 98 kilovolts alternating current to ground or 139 kilovolts direct current to ground, clearances less than those required by Table 235-6 are permitted for systems with known switching surge factor. (See Rule 235E3.)

2. Suspension Insulators

   Where suspension insulators are used and are not restrained from movement, the clearance shall be increased so that the string of insulators may swing transversely throughout a range of insulator swing up to its maximum design swing angle without reducing the values given in Rule 235E1. The maximum design swing angle shall be based on a 6 pounds per square foot (290 Pa) wind on the conductor at final sag at 60 °F (15 °C). This may be reduced to a 4 pounds per square foot (190 Pa) wind in areas sheltered by buildings, terrain, or other obstacles. The displacement of the wires, conductors, and cables shall include deflection of flexible structures and fittings, where such deflection would reduce the clearance.

3. Alternate Clearances for Voltages Exceeding 98 kV Alternating Current to Ground or 139 kV Direct Current to Ground

   The clearances specified in Rules 235E1 and 235E2 may be reduced for circuits with known switching surge factors but shall not be less than the following.

a. Alternate Clearances to Anchor Guys, Surge Protection Wires, and Vertical or Lateral Conductors
The alternate clearances shall be not less than the crossing clearances required by Rule 233B2 and Rules 233C3a and 233C3b for the conductor voltages concerned. For the purpose of this rule, anchor guys and surge protection wires shall be assumed to be at ground potential. The limits of Rule 235E3b(2) shall apply to the clearance derived from Rules 233C3a and 233C3b.

b. Alternate Clearance to Surface of Support Arms and Structures

(1) Alternate Clearance

(a) Basic Computation
The alternate clearances shall be maintained under the expected loading conditions and shall be not less than the electrical clearances computed from the following equation. For convenience, clearances for typical system voltages are shown in Table 235-7.

$$D = 39.37\left[\frac{V \cdot (PU) \cdot a}{500\,K}\right]^{1.667} b \text{ (in)}$$

$$D = 1.00\left[\frac{V \cdot (PU) \cdot a}{500\,K}\right]^{1.667} b \text{ (m)}$$

where

V = maximum alternating-current crest operating voltage to ground or maximum direct current operating voltage to ground in kilovolts;

PU = maximum switching surge factor expressed in per-unit peak voltage to ground and defined as a switching surge level for circuit breakers corresponding to 98% probability that the maximum switching surge generated per breaker operation does not exceed this surge level, or the maximum anticipated switching surge level generated by other means, whichever is greater;

a = 1.15, the allowance for three standard deviations with fixed insulator supports;
= 1.05, the allowance for one standard deviation with free swinging insulators;

b = 1.03, the allowance for nonstandard atmospheric conditions;

K = 1.2, the configuration factor for conductor-to-tower window.

(b) Atmospheric Correction
The value of D shall be increased 3% for each 1000 ft (300 mm) in excess of 1500 ft (450 m) above mean sea level.

(2) Limits
The alternate clearance shall not be less than the clearance of Table 235-6 for 169 kV alternating current. The alternate clearance shall be checked for adequacy of clearance to workmen and increased, if necessary, where work is to be done on the structure while the circuit is energized. (Also see Part 4.)

F. Clearances Between Supply Circuits of Different Voltage Classifications on the Same Support Arm

Supply circuits of any one voltage classification as given in Table 235-5 may be maintained on the same support arm with supply circuits of the next consecutive voltage classification only under one or more of the following conditions:

1. If they occupy positions on opposite sides of the structure.
2. If in bridge-arm or sidearm construction, the clearance is not less than the climbing space required for the higher voltage concerned and provided for in Rule 236.
3. If the higher voltage conductors occupy the outer positions and the lower voltage conductors occupy the inner positions.
4. If series lighting or similar supply circuits are ordinarily dead during periods of work on or above the support arm concerned.
5. If the two circuits concerned are communication circuits used in the operation of supply lines, and supply circuits of less than 8.7 kilovolts, and are owned by the same

utility, provided they are installed as specified in Rule 235F1 or 235F2.

G. Conductor Spacing: Vertical Racks

Conductors or cables may be carried on vertical racks or separate brackets other than wood placed vertically on one side of the structure and securely attached thereto with less clearance between the wires, conductors, or cables than specified in Rule 235C if all the following conditions are met:

1. The voltage shall not be more than 750 volts, except supply cables and conductors meeting Rules 230C1 or 230C2 which may carry any voltage.
2. Conductors shall be of the same material or materials, except that different materials may be used if their sag tension characteristics and arrangement are such that the spacing specified in rule 235G3 is maintained under all service conditions.
3. Vertical spacing between conductors shall be not less than the following:

Span length		Vertical spacing between conductors	
(ft)	(m)	(in)	(mm)
0 to 150	0 to 45	4	100
150 to 200	45 to 60	6	150
200 to 250	60 to 75	8	200
250 to 300	75 to 90	12	300

EXCEPTION: The vertical spacing may be reduced where the conductors are held apart by intermediate spacers, but may not be less than 4 in (100 mm).

Fig 235-1
Clearance Diagram for Energized Conductor

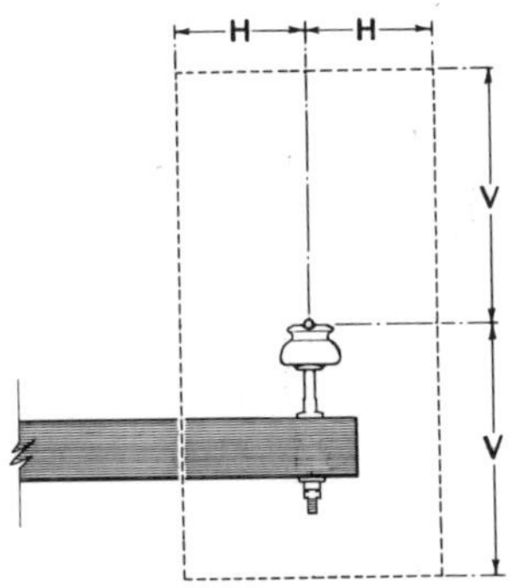

V = Vertical clearance
H = Horizontal clearance

Table 235-1 Horizontal Clearance Between Wires, Conductors, or Cables at Supports

(All voltages are between conductors involved except for railway feeders, which are to ground. See also Rule 235A.)

Class of circuit	Clearance (in)	Clearance (mm)	Notes
Open communication conductors	6	150	Does not apply at conductor transposition points.
	3	75	Permitted where pin spacings less than 6 in (150 mm) have been in regular use. Does not apply at conductor transposition points.
Railway feeders:			Where 10 to 12 in (250 to 300 mm) clearance has already been established by practice, it may be continued, subject to the provisions of Rule 235B1b, for conductors having apparent sags not over 3 ft (900 mm) and for voltages not exceeding 8.7 kV.
0 to 750 V, AWG No 4/0 or larger	6	150	
0 to 750 V, smaller than AWG No 4/0	12	300	
Over 750 V to 8.7 kV	12	300	
Supply conductors of the same circuit:			
0 to 8.7 kV	12	300	
Over 8.7 to 50 kV	12 plus 0.4 per kV over 8.7 kV	300 plus 10 per kV over 8.7 kV	
Above 50 kV	no value specified	no value specified	
Supply conductors of different circuits:			For all voltages above 50 kV, the additional clearance shall be increased 3% for each 1000 ft (300 m) in excess of 3300 ft (1000 m) above mean sea level. All clearances for voltages above 50 kV shall be based on the maximum operating voltage.
0 to 8.7 kV	12	300	
Over 8.7 to 50 kV	12 plus 0.4 per kV over 8.7 kV	300 plus 10 per kV over 8.7 kV	
Over 50 kV to 814 kV	28.5 plus 0.4 per kV over 50 kV	725 plus 10 per kV over 8.7 kV	

Table 235-2 Horizontal Clearances Between Line Conductors Smaller Than AWG No 2 at Supports, Based on Sags IN

Voltage between conductors (kV) \ Sag (in)	36	48	72	96	120	180	240	But not less than①
	Horizontal clearance (in)							
2.4	14.7	20.5	28.7	35.0	40.3	51.2	60.1	12.0
4.16	15.3	21.1	29.3	35.6	40.9	51.8	60.7	12.0
12.47	17.7	23.5	31.7	38.0	43.3	54.2	63.1	13.5
13.2	18.0	23.8	32.0	38.3	43.6	54.5	63.4	13.8
13.8	18.1	23.9	32.1	38.4	43.7	54.6	63.5	14.0
14.4	18.3	24.1	32.3	38.6	43.9	54.8	63.7	14.3
24.94	21.5	27.3	35.5	41.8	47.1	58.0	66.9	18.5
34.5	24.4	30.2	38.4	44.7	50.0	60.9	69.8	22.4
46	27.8	33.6	41.8	48.1	53.4	64.3	73.2	26.9

①Clearance determined by Table 235-1, Rule 235B1a.

NOTE: Clearance = 0.3 in/kV + 7 $\sqrt{(S/3) - 8}$, where S is the sag in inches.

Table 235-3 Horizontal Clearances Between Line Conductors AWG No 2 or Larger at Supports, Based on Sags IN

Voltage between conductors (kV) \ Sag (in)	36	48	72	96	120	180	240	But not less than①
	Horizontal clearance (in)							
2.4	14.6	16.7	20.2	23.3	26.0	31.7	36.5	12.0
4.16	15.1	17.3	20.8	23.8	26.5	32.2	37.0	12.0
12.47	17.6	19.7	23.6	26.3	29.0	34.7	39.5	13.5
13.2	17.8	20.0	23.5	26.5	29.2	34.9	39.7	13.8
13.8	18.0	20.1	23.7	26.7	29.4	35.1	39.9	14.0
14.4	18.2	20.3	23.8	26.9	29.6	35.3	40.1	14.3
24.94	21.3	23.5	27.0	30.0	32.8	38.4	43.2	18.5
34.5	24.2	26.4	29.9	32.9	35.6	41.3	46.1	22.4
46	27.7	29.8	33.3	36.4	39.1	44.8	49.6	26.9

①Clearance determined by Table 235-1, Rule 235B1a.

NOTE: Clearance = 0.3 in/kV + 8 $\sqrt{S/12}$, where S is the sag in inches.

Table 235-2 Horizontal Clearances Between Line Conductors Smaller Than AWG No 2 at Supports, Based on Sags **M**

Sag (in) / Voltage between conductors (kV)	36	48	72	96	120	180	240	But not less than①
	Horizontal clearance (mm)							
2.4	375	520	730	890	1025	1300	1525	300
4.16	390	535	745	905	1040	1315	1540	300
12.47	450	595	805	965	1100	1375	1605	345
13.2	455	605	810	970	1105	1385	1610	350
13.8	460	605	815	975	1110	1385	1615	355
14.4	465	610	820	980	1115	1390	1620	365
24.94	545	695	900	1060	1195	1475	1700	470
34.5	620	765	975	1135	1270	1545	1775	570
46	705	855	1060	1220	1356	1635	1860	685

① Clearance determined by Table 235-1, Rule 235B1a.

NOTE: Clearance = 7.6 mm/kV + 7 $\sqrt{8.5S - 5080}$, where S is the sag in millimeters.

Table 235-3 Horizontal Clearances Between Line Conductors AWG No 2 or Larger at Supports, Based on Sags **M**

Sag (in) / Voltage between conductors (kV)	36	48	72	96	120	180	240	But not less than①
	Horizontal clearance (mm)							
2.4	370	425	515	590	660	805	925	300
4.16	385	440	530	605	675	820	940	300
12.47	445	500	600	670	735	880	1005	345
13.2	450	510	595	675	740	885	1010	350
13.8	455	510	600	680	745	890	1015	355
14.4	460	515	605	685	750	895	1020	365
24.94	540	595	685	760	835	975	1100	470
34.5	615	670	760	835	905	1050	1170	570
46	705	755	845	925	995	1140	1260	685

① Clearance determined by Table 235-1, Rule 235B1a.

NOTE: Clearance = 7.6 mm/kV + 8 $\sqrt{2.12S}$, where S is the sag in millimeters.

Table 235-4. Electrical Clearances in Rule 235B3a(1)
(Add 3% for each 1000 ft (300 m) in excess of 1500 ft (450 m) above mean sea level.)

Maximum operating voltage phase to phase (kV)	Switching surge factor (per unit)	Switching surge (kV)	Electrical component of clearance (ft)	(m)
242	2.6 or less	890 or less	① 6.3	1.90
	2.8	958	7.2	2.20
	3.0	1027	8.1	2.50
	3.2 or more	1095 or more	② 8.8	2.70
362	1.8	893 or less	① 6.4	1.95
	2.0	1024	8.0	2.45
	2.2	1126	9.5	2.90
	2.4	1228	10.9	3.3
	2.6	1330	12.5	3.8
	2.7 or more	1382 or more	② 12.8	3.9
550	1.6	1245	11.2	3.4
	1.8	1399	13.6	4.1
	2.0	1555	16.2	4.9
	2.2	1711	19.0	5.8
	2.3	1789 or more	② 19.1	5.8
800	1.6	1810	20.8	6.3
	1.8	2037	25.3	7.7
	1.9 or more	2149 or more	② 27.4	8.4

① Limited by Rule 235B3b.
② Need not be greater than specified in Rule 235B1 and 235B2.

Table 235-5 Vertical Clearance Between Conductors at Supports

(When using column and row headings, voltages are phase to ground for effectively grounded circuits and those other circuits where all ground faults are cleared by promptly de-energizing the faulted section, both initially and following subsequent breaker operations. See the definitions section for voltages of other systems.)

Conductors usually at upper levels / Conductors usually at lower levels	Supply cables meeting Rule 230C1, 2, or 3; neutral conductors meeting Rule 230E1 (in)	Open supply conductors		
		0 to 8.7 kV (in)	Over 8.7 to 50 kV: Same utility ⑧ (in)	Over 8.7 to 50 kV: Different utilities ⑨ (in)
1. Communication conductors				
a. Ordinary	40 ①⑥	40	40	40 plus 0.4 per kV ⑦ over 8.7 kV
b. Used in operation of supply lines	16	16 ②	40	40 plus 0.4 per kV ⑦ over 8.7 kV
2. Supply conductors				
a. 0 to 750 V; supply cables meeting Rule 230C1, 2, or 3; neutral conductors meeting Rule 230E1	16	16 ③	16 plus 0.4 per kV ⑦ over 8.7 kV	40 plus 0.4 per kV ⑦ over 8.7 kV
b. Over 750 V to 8.7 kV		16 ③	16 plus 0.4 per kV ⑤⑦ over 8.7 kV	40 plus 0.4 per kV ⑦ over 8.7 kV
c. Over 8.7 kV to 22 kV				
(1) If worked on alive with live-line tools and adjacent circuits are neither de-energized nor covered with shields or protectors			16 plus 0.4 per kV ⑦ over 8.7 kV	40 plus 0.4 per kV ⑦ over 8.7 kV

(2) If not worked on alive except when adjacent circuits (either above or below) are de-energized or covered by shields or protectors, or by the use of live-line tools not requiring linemen to go between live wires	16 plus 0.4 per kV ④⑦ over 8.7 kV	16 plus 0.4 per kV ④⑦ over 8.7 kV
d. Exceeding 22 kV, but not exceeding 50 kV	16 plus 0.4 per kV ④⑦ over 8.7 kV	16 plus 0.4 per kV ④⑦ over 8.7 kV

① Where supply circuits of 600 V or less, with transmitted power of 5000 W or less, are run below communication circuits in accordance with Rule 220B2 the clearance may be reduced to 16 in.

② This shall be increased to 40 in when the communication conductors are carried above supply conductors unless the communication-line-conductor size is that required for grade C supply lines.

③ Where conductors are operated by different utilities, a vertical clearance of not less than 40 in is recommended.

④ These values do not apply to conductors of the same circuit or circuits being carried on adjacent conductor supports.

⑤ May be reduced to 16 in where conductors are not worked on alive except when adjacent circuits (either above or below) are de-energized or covered by shields or protectors, or by the use of live-line tools not requiring linemen to go between live wires.

⑥ May be reduced to 30 in for supply neutrals meeting Rule 230E1 and cables meeting Rule 230C1 where the supply neutral or messenger is bonded to the communication messenger.

⑦ The greater of phasor difference or phase-to-ground voltage; see Rule 235A3.

⑧ *Example:* For a 50 kV-to-ground conductor above a 22 kV-to-ground conductor, the required clearance is 16 in + 25 in = 41 in when the conductors are 180° out of phase.

⑨ *Example:* For a 50 kV-to-ground conductor above a 22 kV-to-ground conductor, the required clearance is 40 in + 25 in = 65 in when the conductors are 180° out of phase.

Table 235-5 Vertical Clearance Between Conductors at Supports

M

(When using column and row headings, voltages are phase to ground for effectively grounded circuits and those other circuits where all ground faults are cleared by promptly de-energizing the faulted section, both initially and following subsequent breaker operations. See the definitions section for voltages of other systems.)

Conductors usually at upper levels / Conductors usually at lower levels	Supply cables meeting Rule 230C1, 2, or 3; neutral conductors meeting Rule 230E1 (m)	Open supply conductors		
		0 to 8.7 kV (m)	Over 8.7 to 50 kV: Same utility (8) (m)	Over 8.7 to 50 kV: Different utilities (9) (m)
1. Communication conductors				
a. Ordinary	1.00 (1)(6)	1.00	1.00	1.00 plus 0.01 per kV (7) over 8.7 kV
b. Used in operation of supply lines	0.41	0.41 (2)	1.00	1.00 plus 0.01 per kV (7) over 8.7 kV
2. Supply conductors				
a. 0 to 750 V; supply cables meeting Rule 230C1, 2, or 3; neutral conductors meeting Rule 230E1	0.41	0.41 (3)	0.41 plus 0.01 per kV (7) over 8.7 kV	1.00 plus 0.01 per kV (7) over 8.7 kV
b. Over 750 V to 8.7 kV		0.41 (3)	0.41 plus 0.01 per kV (5)(7) over 8.7 kV	1.00 plus 0.01 per kV (7) over 8.7 kV
c. Over 8.7 kV to 22 kV				
(1) If worked on alive with live-line tools and adjacent circuits are neither de-energized nor covered with shields or protectors			0.41 plus 0.01 per kV (7) over 8.7 kV	1.00 plus 0.01 per kV (7) over 8.7 kV

(2) If not worked on alive except when adjacent circuits (either above or below) are de-energized or covered by shields or protectors, or by the use of live-line tools not requiring linemen to go between live wires	0.41 plus 0.01 per kV ④⑦ over 8.7 kV	0.41 plus 0.01 per kV ④⑦ over 8.7 kV
d. Exceeding 22 kV, but not exceeding 50 kV	0.41 plus 0.01 per kV ④⑦ over 8.7 kV	0.41 plus 0.01 per kV ④⑦ over 8.7 kV

① Where supply circuits of 600 V or less, with transmitted power of 5000 W or less, are run below communication circuits in accordance with Rule 220B2 the clearance may be reduced to 0.41 m.

② This shall be increased to 1.00 m when the communication conductors are carried above supply conductors unless the communication-line-conductor size is that required for grade C supply lines.

③ Where conductors are operated by different utilities, a vertical clearance of not less than 1.00 m is recommended.

④ These values do not apply to conductors of the same circuit or circuits being carried on adjacent conductor supports.

⑤ May be reduced to 0.41 m where conductors are not worked on alive except when adjacent circuits (either above or below) are de-energized or covered by shields or protectors, or by the use of live-line tools not requiring linemen to go between live wires.

⑥ May be reduced to 0.75 m for supply neutrals meeting Rule 230E1 and cables meeting Rule 230C1 where the supply neutral or messenger is bonded to the communication messenger.

⑦ The greater of phasor difference or phase-to-ground voltage; see Rule 235A3.

⑧ *Example:* For a 50 kV-to-ground conductor above a 22 kV-to-ground conductor, the required clearance is 0.41 m + 0.64 m = 1.05 m when the conductors are 180° out of phase.

⑨ *Example:* For a 50 kV-to-ground conductor above a 22 kV-to-ground conductor, the required clearance is 1.00 m + 0.64 m = 1.64m when the conductors are 180° out of phase.

Table 235-6 Clearance in Any Direction from Line Conductors to Supports and to Vertical or Lateral Conductors, Span, or Guy Wires Attached to the Same Support

	Communication lines		Supply lines		
			Circuit phase-to-phase voltage		
Clearance of line conductors from	In general (in)	On jointly used structures (in)	0 to 8.7 kV (in)	Over 8.7 to 50 kV (in)	Over 50 to 814 kV ④⑨ (in)
1. Vertical and lateral conductors:					
a. Of the same circuit	3	3	3	3 plus 0.25 per kV over 8.7 kV	no value specified
b. Of other circuits ⑫	3	3	⑤6	6 plus 0.4 per kV over 8.7 kV	23 plus 0.4 per kV over 50 kV
2. Span or guy wires, ⑪ or messengers attached to same structure:					
a. When parallel to line	⑦3	①⑦6	①12	12 plus 0.4 per kV over 8.7 kV	29 plus 0.4 per kV over 50 kV
b. Anchor guys	⑦3	①⑦6	①6	6 plus 0.25 per kV over 8.7 kV	16 plus 0.25 per kV over 50 kV
c. All other	⑦3	①⑦6	6	6 plus 0.4 per kV over 8.7 kV	23 plus 0.4 per kV over 50 kV

3. Surface of support arms	②3	②3	⑥⑧3	3 plus 0.2 per kV over 8.7 kV⑥⑧⑩	11 plus 0.2 per kV over 50 kV
4. Surface of structures:					
a. On jointly used structures	—	②5	③⑥⑧5	5 plus 0.2 per kV over 8.7 kV ⑥⑧⑩	13 plus 0.2 per kV over 50 kV
b. All other	②3	—	⑥⑧3	3 plus 0.2 per kV over 8.7 kV ⑥⑧⑩	11 plus 0.2 per kV over 50 kV

① For guy wires, if practical. For clearances between span wires and communication conductors, see Rule 238C.

On jointly used structures, guys which pass within 12 in of supply conductors, and also pass within 12 in of communication cables, shall be protected with a suitable insulating covering where the guy passes the supply conductors, unless the guy is effectively grounded or insulated with a strain insulator at a point below the lowest supply conductor and above the highest communication cable.

The clearance from an insulated or effectively grounded guy to a communication cable may be reduced to 3 inches when abrasion protection is provided on the guy or communication cable.

② Communication conductors may be attached to supports on the sides or bottom of crossarms or surfaces of poles with less clearances.

③ This clearance applies only to supply conductors at the support below communication conductors, on jointly used structures.

Where supply conductors are above communication conductors, this clearance may be reduced to 3 in.

④ All clearances for line over 50 kV shall be based on the maximum operating voltage. For voltages exceeding 814 kV, the clearance shall be determined by the alternate method given by Rule 235E3.

⑤ For supply circuits of 0 to 750 V, this clearance may be reduced to 3 in.

⑥ A neutral conductor meeting Rule 230E1 may be attached directly to the structure surface.

(continued on next page)

Footnotes for Table 235-6 *(Continued)*

IN

⑦ Guys and messengers may be attached to the same strain plates or to the same through bolts.

⑧For open supply circuits of 0 to 750 V and supply cables of all voltages meeting Rule 230C1, 2 or 3, this clearance may be reduced to 1 in. No clearance is specified for phase conductors of such cables where they are physically restrained by a suitable bracket from abrasion against the pole.

⑨ The additional clearance for voltages in excess of 50 kV specified in Table 235-6 shall be increased 3% for each 1000 ft in excess of 3300 ft above mean sea level.

⑩ Where the circuit is effectively grounded and the neutral conductor meets Rule 230E1, phase-to-neutral voltage shall be used to determine the clearance from the surface of support arms and structures.

⑪ These clearances may be reduced by not more than 25% to a guy insulator, provided that full clearance is maintained to its metallic end fittings and the guy wires. The clearance to an insulated section of a guy between two insulators may be reduced by not more than 25% provided that full clearance is maintained to the uninsulated portion of the guy.

⑫ Phase-to-phase voltages shall be determined according to Rule 235A3.

Table 235-6 Clearance in Any Direction from Line Conductors to Supports and to Vertical or Lateral Conductors, Span, or Guy Wires Attached to the Same Support

M

Clearance of line conductors from	Communication lines		Supply lines		
			Circuit phase-to-phase voltage		
	In general (mm)	On jointly used structures (mm)	0 to 8.7 kV (mm)	Over 8.7 to 50 kV (mm)	Over 50 to 814 kV ④⑨ (mm)
1. Vertical and lateral conductors:					
a. Of the same circuit	75	75	75	75 plus 6.5 per kV over 8.7 kV	no value specified

b. Of other circuits ⑫	75	75	⑤150	150 plus 10 per kV over 8.7 kV	580 plus 10 per kV over 50 kV
2. Span or guy wires, ⑪ or messengers attached to same structure:					
a. When parallel to line	⑦ 75	①⑦150	① 300	300 plus 10 per kV over 8.7 kV	740 plus 10 per kV over 50 kV
b. Anchor guys	⑦ 75	①⑦150	① 150	150 plus 6.5 per kV over 8.7 kV	410 plus 6.5 per kV over 50 kV
c. All other	⑦ 75	①⑦150	150	150 plus 10 per kV over 8.7 kV	580 plus 10 per kV over 50 kV
3. Surface of support arms	②75	② 75	⑥⑧75	75 plus 5 per kV over 8.7 kV ⑥⑧⑩	280 plus 5 per kV over 50 kV
4. Surface of structures:					
a. On jointly used structures	—	② 125	③⑥⑧ 125	125 plus 5 per kV over 8.7 kV ⑥⑧⑩	330 plus 5 per kV over 50 kV
b. All other	②75	—	⑥⑧75	75 plus 5 per kV over 8.7 kV ⑥⑧⑩	280 plus 5 per kV over 50 kV

① For guy wires, if practical. For clearances between span wires and communication conductors, see Rule 238C.

On jointly used structures, guys which pass within 300 mm of supply conductors, and also pass within 300 mm of communication cables, shall be protected with

(continued on next page)

Footnotes for Table 235-6 *(Continued)*

a suitable insulating covering where the guy passes the supply conductors, unless the guy is effectively grounded or insulated with a strain insulator at a point below the lowest supply conductor and above the highest communication cable.

The clearance from an insulated or effectively grounded guy to a communication cable may be reduced to 75 mm when abrasion protection is provided on the guy or communication cable.

② Communication conductors may be attached to supports on the sides or bottom of crossarms or surfaces of poles with less clearances.

③ This clearance applies only to supply conductors at the support below communication conductors, on jointly used structures.

Where supply conductors are above communication conductors, this clearance may be reduced to 75 mm.

④ All clearances for line over 50 kV shall be based on the maximum operating voltage. For voltages exceeding 814 kV, the clearance shall be determined by the alternate method given by Rule 235E3.

⑤ For supply circuits of 0 to 750 V, this clearance may be reduced to 75 mm.

⑥ A neutral conductor meeting Rule 230E1 may be attached directly to the structure surface.

⑦ Guys and messengers may be attached to the same strain plates or to the same through bolts.

⑧ For open supply circuits of 0 to 750 V and supply cables of all voltages meeting Rule 230C1, 2 or 3, this clearance may be reduced to 1 in. No clearance is specified for phase conductors of such cables where they are physically restrained by a suitable bracket from abrasion against the pole.

⑨ The additional clearance for voltages in excess of 50 kV specified in Table 235-6 shall be increased 3% for each 300 m in excess of 1000 m above mean sea level.

⑩ Where the circuit is effectively grounded and the neutral conductor meets Rule 230E1, phase-to-neutral voltage shall be used to determine the clearance from the surface of support arms and structures.

⑪ These clearances may be reduced by not more than 25% to a guy insulator, provided that full clearance is maintained to its metallic end fittings and the guy wires. The clearance to an insulated section of a guy between two insulators may be reduced by not more than 25% provided that full clearance is maintained to the uninsulated portion of the guy.

⑫ Phase-to-phase voltages shall be determined according to Rule 235A3.

Table 235-7 Clearance in Any Direction from Line Conductors to Supports

Maximum operating voltage phase to phase (kV)	Switching surge factor (per unit)	Switching surge (kV)	Computed clearance to supports: Fixed (in)	Fixed (m)	Free swinging at maximum angle (in)	Free swinging at maximum angle (m)
242	2.4	474	① 35	0.89	① 35	0.89
	2.6	514	40	1.00	① 35	0.89
	2.8	553	45	1.14	38	0.97
	3.0	593	② 50	1.27	43	1.10
	3.2	632	② 50	1.27	48	1.20
362	1.6	473	① 35	0.89	① 35	0.89
	1.8	532	42	1.07	36	0.90
	2.0	591	50	1.27	48	1.20
	2.2	650	59	1.50	51	1.30
	2.4	709	68	1.73	59	1.50
	2.5	739	② 73	1.85	63	1.60
550	1.6	719	70	1.80	60	1.50
	1.8	808	85	2.15	73	1.85
	2.0	898	101	2.60	87	2.20
	2.2	988	② 111	2.80	101	2.60
800	1.6	1045	130	3.3	111	2.80
	1.8	1176	158	4.0	135	3.4
	1.9	1241	② 161	4.1	148	3.8
	2.0	1306	② 161	4.1	② 161	4.1

① Limited by Rule 235E3(b)(2).
② Need not be greater than specified in Rules 235E1 and 2.

236. Climbing Space

The following requirements apply only to portions of structures which workers ascend.

A. Location and Dimensions

1. A climbing space having the horizontal dimensions specified in Rule 236E shall be provided past any conductors, support arms, or other parts.
2. The climbing space need be provided on one side or corner of the support only.
3. The climbing space shall extend vertically past any conductor or other part between levels above and below the conductor as specified in Rules 236E, F, G, and I, but may otherwise be shifted from any side or corner of the support to any other side or corner.

B. Portions of Supporting Structures in Climbing Space

Portions of the supporting structure when included in one side or corner of the climbing space are not considered to obstruct the climbing space.

C. Support Arm Location Relative to Climbing Space

RECOMMENDATION: Support arms should be located on the same side of the pole.

EXCEPTION: This recommendation does not apply where double crossarms are used on any pole or where crossarms on any pole are not all parallel.

D. Location of Equipment Relative to Climbing Space

All supply and communication equipment such as transformers, regulators, capacitors, cable terminals (potheads), amplifiers, loading coils, surge arresters, switches, etc, when located below conductors or other attachments shall be mounted outside of the climbing space.

E. Climbing Space Between Conductors

Climbing space between conductors shall be not less than the horizontal dimensions specified in Table 236-1. These dimensions are intended to provide a clear climbing space of 24 in (600 mm) while the conductors bounding the climbing space are covered with temporarily installed protective covering rated for the voltage involved. The climbing space shall be provided both along and across the line, and shall be projected vertically not less than 40 in (1.00 mm) above and below the limiting conductors. Where communication conductors are above supply conductors of more than

8.7 kilovolts to ground or 15 kilovolts line to line, the climbing space shall be projected vertically at least 60 in (1.50 m) above the highest supply conductors.

EXCEPTION 1: This rule does not apply if it is the unvarying practice of the employers concerned to prohibit employees from ascending beyond the conductors or equipment of a given line or structure unless the conductors or equipment are de-energized.

EXCEPTION 2: For supply conductors carried on a structure in a position below communication facilities in the manner permitted in Rule 220B2, the climbing space need not extend more than 2 ft (600 mm) above such supply space.

EXCEPTION 3: If the conductors are owned, operated, or maintained by the same utility, the climbing space may be provided by temporarily moving the line conductors using live line tools.

F. Climbing Space on Buckarm Construction

Method of Providing Climbing Space on Buckarm Construction

The full width of climbing space shall be maintained on buckarm construction and shall extend vertically in the same position at least 40 in (1.00 m) [or 60 in (1.50 m) where required by Rule 236E] above and below any limiting conductor.

With circuits of less than 8.7 kilovolts to ground or 15 kilovolts line to line and span lengths not exceeding 150 ft (45 m) and sags not exceeding 15 in (380 mm) for wires of AWG No 2 and larger sizes, or 30 in (750 mm) for wires smaller than AWG No 2 and larger sizes, or 30 in (750 mm) for wires smaller than AWG No 2, a six-pin crossarm having pin spacing of 14 1/2 in (370 mm) may be used to provide a 30 in (750 mm) climbing space on one corner of a junction pole by omitting the pole pins on all arms, and inserting pins midway between the remaining pins so as to give a spacing of 7 1/4 in (185 mm), provided that each conductor on the end of every arm is tied to the same side of its insulator, and that the spacing on the next pole is not less than 14 1/2 in (370 mm).

G. Climbing Space Past Longitudinal Runs Not on Support Arms

The full width of climbing space shall be provided past longitudinal runs and shall extend vertically in the same position from 40 in (1.00 m) below the run to a point 40 in (1.00 m) above [or 60 in (1.50 m) where required by Rule

236E]. The width of climbing space shall be measured from the longitudinal run concerned. Longitudinal runs on racks, or cables on messengers, are not considered as obstructing the climbing space if all wires concerned are covered by rubber protective equipment or otherwise guarded as an unvarying practice before workers climb past them. This does not apply where communication conductors are above the longitudinal runs concerned.

EXCEPTION 1: If a supply longitudinal run is placed on the side or corner of the supporting structure where climbing space is provided, the width of climbing space shall be measured horizontally from the center of the structure to the nearest supply conductors on support arms, under both of the following conditions:

(i) Where the longitudinal run consists of open supply conductors carrying not more than 750 V, or supply cables and conductors meeting Rule 230C, all voltages; and is supported close to the structure as by brackets, racks, or pins close to the structure.

(ii) Where the nearest supply conductors on support arms are parallel to and on the same side of the structure as the longitudinal run and within 4 ft (1.20 m) above or below the run.

EXCEPTION 2: For supply conductors carried on a structure in a position below communication facilities in the manner permitted in Rule 220B2, the climbing space need not extend more than 2 ft (600 mm) above such supply space.

EXCEPTION 3: A service drop less than 750 volts and meeting Rule 230C is not considered to obstruct the climbing space if all conductors concerned are covered by rubber protective equipment or otherwise guarded as an unvarying practice before workers climb past them, provided that such a service drop is (1) not closer to the longitudinal run at the point of attachment than the diameter of the pole plus 5 in (125 mm) measured horizontally and (2) not closer than 38 in (950 mm) measured horizontally to the longitudinal run at a point 30 in (750 mm) on the run measured from the point of attachment at the pole. See Fig 236-1.

H. Climbing Space Past Vertical Conductors

Vertical runs physically protected by suitable conduit or other protective covering and securely attached without spacers to the surface of the line structure are not considered to obstruct the climbing space.

I. Climbing Space Near Ridge-Pin Conductors

The climbing space specified in Table 236-1 shall be provided above the top support arm to the ridge-pin conductor but need not be carried past it.

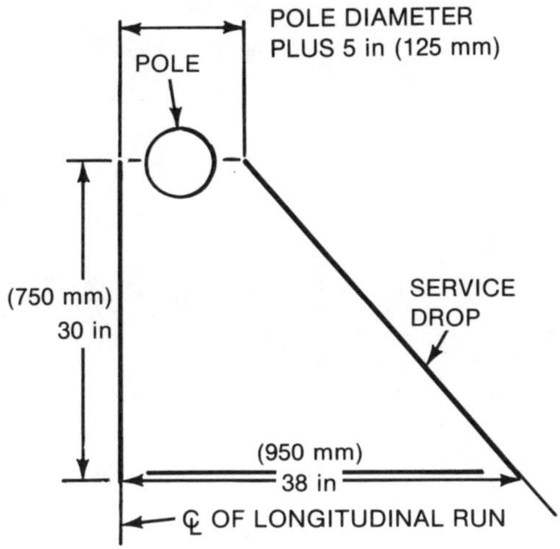

LEGEND

——— BOUNDARY OF CLIMBING SPACE

Fig 236-1
Rule 236G, Exception 3

Table 236-1 Horizontal Clearance Between Conductors Bounding the Climbing Space

(All voltages are between the two conductors bounding the climbing space except for communications conductors which are voltage to ground. Where the two conductors are in different circuits, the voltage between conductors shall be the arithmetic sum of the voltages of each conductor to ground for a grounded circuit or phase to phase for an ungrounded circuit.)

Character of conductors adjacent to climbing space	Voltage of conductors	Horizontal clearance between conductors bounding the climbing space ③							
		On structures used solely by				On jointly used structures			
		Communication conductors		Supply conductors		Supply conductors above communication conductors		Communication conductors above supply conductors ①	
		(in)	(m)	(in)	(m)	(in)	(m)	(in)	(m)
1. Communication conductors	0 to 150 V	no requirements		—		②		no requirements	
	exceeding 150 V	24 recommended	0.60	—		②		24 recommended	0.60
2. Supply cables meeting Rule 230C1	all voltages			—		②		no requirements	
3. Supply cables meeting Rule 230C2 or 3	all voltages	—	—	24	0.60	24	0.60	30	0.75

4.	Open supply line conductors and supply cables meeting Rule 230D	0 to 750 V	—	—	24	0.60	24	0.60	30	0.75
		750 V to 15 kV	—	—	30	0.75	30	0.75	30	0.75
		15 kV to 28 kV	—	—	36	0.90	36	0.90	36	0.90
		28 kV to 38 kV	—	—	40	1.00	40	1.00		
		38 kV to 50 kV	—	—	46	1.17	46	1.17		
		50 kV to 73 kV	—	—	54	1.40	54	1.40		
		exceeding 73 kV	—	—	>54	>1.40				

① This relation of levels is not, in general, desirable and should be avoided.

② Climbing space shall be the same as required for the supply conductors immediately above, with a maximum of 30 in (0.75 m) except that a climbing space of 16 in (0.41 m) across the line may be employed for communication cables or conductors where the only supply conductors at a higher level are secondaries (0 to 750 V) supplying airport or airway marker lights or crossing over the communication line and attached to the pole top or to a pole top extension fixture.

③ Attention is called to the operating requirements of Rules 441A and 446C, Part 4, of this code.

237. Working Space

A. Location of Working Spaces

Working spaces shall be provided on the climbing face of the structure at each side of the climbing space.

B. Dimensions of Working Spaces

1. Along the Support Arm

   The working space shall extend from the climbing space to the outmost conductor position on the support arm.

2. At Right Angles to the Support Arm

   The working space shall have the same dimension as the climbing space (see Rule 236E). This dimension shall be measured horizontally from the face of the support arm.

3. Vertically

   The working space shall have a height not less than that required by Rule 235 for the vertical separation of line conductors carried at different levels on the same support.

C. Location of Vertical and Lateral Conductors Relative to Working Spaces

The working spaces shall not be obstructed by vertical or lateral conductors. Such conductors shall be located on the opposite side of the pole from the climbing side or on the climbing side of the pole at a distance from the support arm at least as great as the width of climbing space required for the highest voltage conductors concerned. Vertical conductors enclosed in suitable conduit may be attached on the climbing side of the structure.

D. Location of Buckarms Relative to Working Spaces

Buckarms may be used under any of the following conditions, provided the climbing space is maintained. Climbing space may be obtained as in Rule 236F.

1. Standard Height of Working Space

   Lateral working space of the height required by Table 235-5 shall be provided between the lateral conductors attached to the buckarm and the line conductors. This may be accomplished by increasing the spacing between the line support arms, as shown in Fig 237-1.

2. Reduced Height of Working Space

   Where no circuits exceeding 8.7 kilovolts to ground or 15 kilovolts line to line are involved and the clearances of

Rules 235B1a and 235B1b are maintained, conductors supported on buckarms may be placed between line conductors having normal vertical spacing, even though such buckarms obstruct the normal working space, provided that a working space of not less than 18 in (450 mm) in height is maintained either above or below line conductors and buckarm conductors.

EXCEPTION: The above working space may be reduced to 12 in (300 mm) if both of the following conditions exist:

(i) Not more than two sets of the line arms and buckarms are involved.

(ii) Working conditions are rendered safe by providing rubber protective equipment or other suitable devices to insulate and cover line conductors and equipment which are not being worked upon.

E. Guarding of Energized Equipment

Exposed energized parts of equipment such as switches, circuit breakers, surge arresters, etc., shall be enclosed or guarded if all of the following conditions apply:

1. The equipment is located below the top conductor support.
2. The equipment is located on the climbing side of the structure.
3. The requirements of Rule 441A, Part 4, of this code cannot be met.

F. Working Clearances from Energized Equipment

All parts of equipment such as switches, fuses, transformers, surge arresters, etc, or other connections which may require operation or adjustment while energized and exposed at such times, shall be so arranged with respect to each other, other equipment, vertical and lateral conductors, and portions of the supporting structure, including supporting platforms or structural members, that in adjustment or operation no portion of the body, including the hands, need be brought closer to any exposed energized parts or conductors than permitted in Part 4, Rules 441A or 446C, of this code.

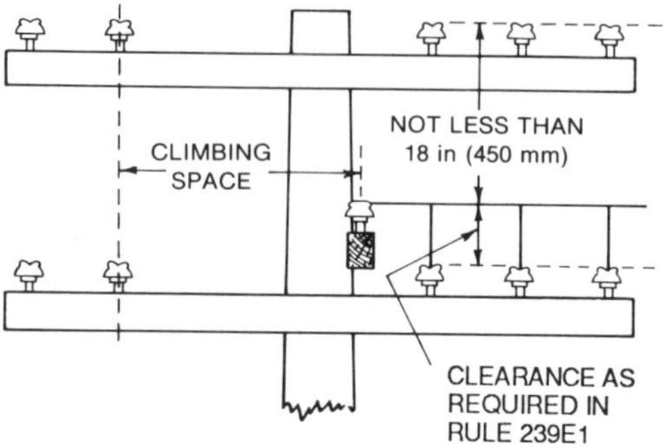

Fig 237-1
Obstruction of Working Space by Buckarm

238. Vertical Clearance Between Certain Communication and Supply Facilities Located on the Same Structure

A. Equipment

For the purpose of measuring clearances under this rule, *equipment* shall be taken to mean noncurrent-carrying metal parts of equipment, including metal supports for cables or conductors, and metal support braces which are attached to metal supports or are less than 1 in (25 mm) from transformer cases or hangers which are not effectively grounded.

B. Clearances in General

Vertical clearances between supply conductors and communication equipment, between communication conductors and supply equipment, and between supply and communication equipment shall be as specified in Table 238-1 except as provided in Rule 238C.

C. Clearances for Span Wires or Brackets

Span wires or brackets carrying luminaires, traffic signals, or trolley conductors shall have at least the vertical clearances in inches from communication equipment set forth in Table 238-2.

D. Clearance of Drip Loops of Luminaire or Traffic Signal Brackets

If a drip loop of conductors entering a luminaire bracket or traffic signal bracket from the surface of the structure is above a communication cable, the lowest point of the loop shall be at least 12 in (300 mm) above communication cable or through bolt.

EXCEPTION: The above clearance may be reduced to 3 in (75 mm) if the loop is covered by a suitable non-metallic covering which extends at least 2 in (50 mm) beyond the loop.

Table 238-1. Vertical Clearance Between Supply Conductors and Communication Equipment, Between Communication Conductors and Supply Equipment, and Between Supply and Communication Equipment

(Voltages are phase to ground for effectively grounded circuits and those other circuits where all ground faults are cleared by promptly de-energizing the faulted section, both initially and following subsequent breaker operations. See the definition section for voltages of other systems.)

Supply voltage (kV)	Vertical clearance (in)	(m)
1. Grounded conductor and messenger hardware and supports	30	0.75
2. 0 to 8.7	① 40	1.00
3. Over 8.7	① 40 plus 0.4 per kV over 8.7 kV	1.00 plus 0.01 per kV over 8.7 kV

① Where noncurrent-carrying parts of equipment are effectively grounded consistently throughout well-defined areas and where communication is at lower levels, clearances may be reduced to 30 in (0.75 m).

Table 238-2 Vertical Clearance of Span Wires and Brackets from Communications Lines

	Carrying luminaires or traffic signals				Carrying trolley conductors			
	Not effectively grounded		Effectively grounded		Not effectively grounded		Effectively grounded	
	(in)	(mm)	(in)	(mm)	(in)	(mm)	(in)	(mm)
Above communication support arms	① 20	500	① 20	500	① 20	500	① 20	500
Below communication support arms	③ 40	1000	24	600	24	600	24	600
Above messengers carrying communication cables	① 20	500	4	100	12	300	4	100
Below messengers carrying communication cables	④ 40	1000	4	100	12	300	4	100
From terminal box of communication cable	① 20	500	4	100	② 12	300	4	100
From communication brackets, bridle wire rings, or drive hooks	① 16	410	4	100	4	100	4	100

① This may be reduced to 12 in (300 mm) for either span wires or metal parts of brackets at points 40 in (1.00 m) or more from the structure surface.

② Where it is not practical to obtain a clearance of 1 ft (300 mm) from terminal boxes of communication cables, all metal parts of terminals shall have the greatest possible separation from fixtures or span wires including all supporting screws and bolts of both attachments.

③ This may be reduced to 24 in (600 mm) for luminaires and traffic signals operating at less than 150 V to ground.

④ This may be reduced to 20 in (500 mm) for luminaires and traffic signals operating at less than 150 V to ground.

239. Clearance of Vertical and Lateral Facilities from Other Facilities and Surfaces on the Same Support

Vertical and lateral conductors shall have the clearances and separations required by this rule from other facilities or surfaces on the same support.

A. General

1. Grounding conductors, neutral conductors meeting Rule 230E1, supply cables meeting Rule 230C1, or conduits enclosing conductors may be placed directly on the support.
2. Supply circuits of the same or next voltage classification may be placed in the same duct, if each circuit or set of wires is enclosed in a metal sheath.
3. Paired communication conductors in rings may be attached directly to a structure or messenger.
4. Insulated supply circuits of 600 volts or less and not exceeding 5000 W may be placed in the same cable with control circuits with which they are associated.

B. Location of Vertical or Lateral Conductors Relative to Climbing Spaces, Working Spaces, and Pole Steps

Vertical or lateral conductors shall be located so that they do not obstruct climbing spaces, or lateral working spaces between line conductors at different levels, or interfere with the safe use of pole steps.

EXCEPTION 1: This rule does not apply to portions of the structure which workers do not ascend while the conductors in question are alive.

C. Conductors Not in Conduit

Conductors not encased in conduit shall have the same clearances from conduits as from other surfaces of structures.

D. Mechanical Protection Near Ground

Where within 8 ft (2.45 m) of the ground, all vertical conductors, cables, and grounding wires shall be protected by a covering which gives suitable mechanical protection.

EXCEPTION 1: This covering may be omitted from armored cables or cables installed in a grounded metal conduit.

EXCEPTION 2: This covering may be omitted from lead-sheathed cables used in rural districts.

EXCEPTION 3: This covering may be omitted from vertical runs of communication cables or conductors.

EXCEPTION 4: This covering may be omitted from grounding wires used in rural districts or in any area where the grounding wire is one of a number of grounding wires used to provide multiple grounds.

EXCEPTION 5: This covering may be omitted from wires which are used solely to protect poles from lightning.

NOTE: See Rule 93D for grounding conductors.

E. Requirements for Vertical and Lateral Supply Conductors on Supply Line Structures or Within Supply Space on Jointly Used Structures

1. General Clearances

   In general, clearances shall be not less than the values specified in Table 239-1 or Rule 235E.

2. Special Cases

   The following requirements apply only to portions of a structure which workers ascend while the conductors in question are alive.

   a. Sidearm Construction

      Vertical conductors in cables meeting Rule 230C1 and grounding wires may be run without insulating protection from supply line conductors on structures used only for supply lines and employing sidearm construction on the side of the structure opposite to the line conductors if climbing space is provided on the line-conductor side of the structure.

   b. Conductors to Luminaires

      On structures used only for supply lines, open wires may be run from the supply line arm directly to the head of a luminaire, provided the clearances of Table 239-1 are obtained and the open wires are substantially supported at both ends.

   c. Conductors of Less Than 300 Volts

      Vertical or lateral secondary supply conductors of not more than 300 volts to ground may be run in multiple-conductor cable attached directly to the structure surface or to support arms in such a manner as to avoid abrasion at the point of attachment. Each conductor of such cable which is not effectively grounded, or the entire cable assembly, shall have an insulating covering required for a conductor of at least 600 volts.

d. Other Conditions

If open wire conductors are within 4 ft (1.20 m) of the pole, vertical conductors shall be run in one of the following ways.

(1) The clearance between open vertical conductors and pole center shall be not less than that given in Table 239-2 within the zone specified in the table.

(2) Within the zone above and below open supply conductor as given in Table 239-2, vertical and lateral conductors may be enclosed in nonmetallic conduit, or in cable protected by an insulating covering, and may be run on the pole surface.

(3) Supply grounding conductors may be run on the pole surface without molding except as required by Rule 239D for mechanical protection near the ground.

F. Requirements for Vertical and Lateral Communications Conductors on Communication Line Structures or Within the Communication Space on Jointly Used Structures

1. Clearances from Wires

The clearances of uninsulated vertical and lateral conductors from other conductors (except those in the same ring run) and from guy, span, or messenger wires shall be 3 in (75 mm).

2. Clearances from Supporting Structure Surfaces

Vertical and lateral insulated communication conductors may be attached directly to a structure. They shall have a vertical clearance of at least 40 in (1.00 m) from any supply conductors (other than vertical runs or luminaire leads) of 8.7 kilovolts or less, or 60 in (1.50 m) if more than 8.7 kilovolts.

EXCEPTION: These clearances do not apply where the supply circuits involved are those carried in the manner specified in Rule 220B2.

G. Requirements for Vertical Supply Conductors Passing Through Communication Space on Jointly Used Line Structures

1. Cables Meeting Rule 230C1

   Cables meeting Rule 230C1 may be fastened directly to the surface of the line structure. Such cables shall be protected with suitable non-metallic covering when passing through the climbing space of all line structures and on those structures carrying trolley attachments or when an ungrounded luminaire is attached below the communication cable. The cable shall be protected with non-metallic covering to a point 40 in (1.0 m) above the highest communication wire and 6 ft (1.8 m) below the lowest trolley attachment or ungrounded luminaire fixture.

2. Jacketed Multiple-Conductor Cables

   Jacketed multiple-conductor cables operating at voltages not exceeding 300 volts to ground may be attached directly to the surface of the line structure. Each conductor shall be insulated for a potential of at least 600 volts. Where used as aerial services, the point where such cables leave the structure shall be at least 40 in (1.00 m) above the highest or 40 in (1.00 m) below the lowest communication attachment. All splices and connections in the cable shall be insulated. No additional protection is required.

3. Cables and Conductors in Conduit or Covering

   Cables and conductors of all voltages may be run in a nonmetallic covering or in a grounded metallic conduit or covering. Where a metallic conduit or covering is not bonded to grounded communication facilities at that structure, such metal conduit or covering shall be protected with a nonmetallic covering under the same conditions and to the same extent as required for cables meeting Rule 230C1 in Rule 239G1.

4. Suspended from Supply Support Arm

   Lamp leads of lighting circuits may be run from supply support arms directly to a bracket or luminaire under the following conditions:

a. The vertical run shall consist of paired wires or multiple-conductor cable securely attached at both ends to suitable brackets and insulators.

b. The vertical run shall be held taut at least 40 in (1.00 m) from the surface of the pole through the communication space at least 12 in (300 mm) beyond the end of any communication support arm by which it passes, and at least 6 in (150 mm) from communication drop wires, and at least 20 in (500 mm) from any communication cable.

c. Insulators attached to luminaire brackets for supporting the vertical run shall be capable of meeting, in the position in which they are installed, the same flashover requirements as the luminaire insulators.

d. Each conductor of the vertical run shall be AWG No 10 or larger.

5. Supply Grounding Conductors

a. Supply grounding conductors may be run bare where there are no trolley attachments or ungrounded street lighting fixtures, or both, located below the communication attachment provided:

(1) The grounding conductor is directly (metallically) connected to a conductor which forms part of an effective grounding system, and

(2) The grounding conductor has no connection to supply equipment between the grounding electrode and the effectively grounded conductor unless the supply equipment has additional connections to the effectively grounded conductor.

b. Supply grounding conductors not conforming to Rule 239G5a shall be protected with a suitable nonmetallic covering to the same extent as required for grounded metal-sheathed cables in Rule 239G1.

6. Clearance from Through Bolts

Vertical runs of supply conductors or cables shall have a clearance of not less than 2 in (50 mm) from exposed through bolts and other exposed metal objects attached thereto which are associated with communication line equipment.

EXCEPTION: Vertical runs of effectively grounded supply conductors may have a clearance of 1 in (25 mm) from the end of the exposed communication through bolts.

7. Multiple Conductor Cables

   Multiple conductor cables operating at voltages not exceeding 600 V between conductors may be attached directly to the surface of the line structure if protected by nonmetallic covering. Each conductor shall be insulated for a potential of at least 600 V. Where used as aerial services, the point where such cables leave the structure shall be at least 40 in (1.00 m) above the highest or 40 in (1.00 m) below the lowest communications attachment. All splices and connections in the cable shall be insulated.

H. Requirements for Vertical Communication Conductors Passing Through Supply Space on Jointly Used Structures

All vertical runs of communication conductors passing through supply space shall be installed as follows.

1. Metal-Sheathed Communication Cables

   Vertical runs of metal-sheath communication cables shall be covered with wood molding, or other suitable nonmetallic material, where they pass trolley feeders or other supply line conductors. This nonmetallic covering shall extend from a point 40 in (1.00 m) above the highest trolley feeders or other supply conductors, to a point 6 ft (1.80 m) below the lowest trolley feeders or other supply conductors, but need not extend below the top of any mechanical protection which may be provided near the ground.

   EXCEPTION: Communication cables may be run vertically on the pole through space occupied by railroad signal supply circuits in the lower position, as permitted in Rule 220B2, without nonmetallic covering within the supply space.

2. Communication Conductors

   Vertical runs of insulated communication conductors shall be covered with wood molding, or other suitable nonmetallic material, to the extent required for metal-sheathed communication cables in Rule 239H1, where such conductors pass trolley feeders or supply conductors.

   EXCEPTION: Communication conductors may be run vertically on the structure through space occupied by railroad-signal supply circuits in the lower position, as permitted in Rule 220B2, without metallic covering within the supply space.

3. Communication Grounding Conductors

Vertical communication grounding conductors shall be covered with wood molding or other nonmetallic material between points at least 6 ft (1.80 m) below and 40 in (1.00 m) above any trolley feeders or other supply line conductors by which they pass.

EXCEPTION: Communication grounding conductors may be run vertically on the structure through space occupied by railroad-signal supply circuits in the lower position, as permitted in Rule 220B2, without nonmetallic covering within the supply space.

4. Separation from Through Bolts

Vertical runs of communication conductors shall have a clearance of one-eighth of the pole circumference but not less than 2 in (50 mm) from through bolts and other metal objects attached thereto which are associated with supply line equipment.

EXCEPTION: Vertical runs of effectively grounded communications conductors may have a separation of 1 in (25 mm) from the end of supply through bolts.

I. Operating Rods

Effectively grounded or insulated operating rods of switches are permitted to pass through the communication space, but shall be located outside of the climbing space.

Table 239-1 Clearance of Vertical and Lateral Conductors IN
(Circuit Phase-to-Phase Voltage)

Clearance of vertical and lateral conductors	0 to 8.7 kV (in)	Over 8.7 to 50 kV (in)	Over 50 kV④ (in)
From surfaces of supports	①②3	3 plus 0.2 per kV over 8.7 kV	11 plus 0.2 per kV over 50 kV
From span, guy, and messenger wires	⑤6	6 plus 0.4 per kV over 8.7 kV③	23 plus 0.4 per kV over 50 kV③

① A neutral conductor meeting Rule 230E1 may be attached directly to the structure surface.

② For supply circuits of 0 to 750 V this clearance may be reduced to 1 in.

③ Multiplier may be reduced to 0.25 in/kV for anchor guys.

④ The additional clearance for voltages in excess of 50 kV specified in Table 239-1, shall be increased 3% for each 1000 ft in excess of 3300 ft above mean sea level.

⑤ For cables meeting Rule 230C3 and operating at 0-750 V, this may be reduced to 2 inches.

Table 239-2 Clearance Between Open Vertical Conductors and Pole Center IN
(Voltages are phase to ground for effectively grounded circuits and those other circuits where all ground faults are cleared by promptly de-energizing the faulted section, both initially and following subsequent breaker operations. See the definition section for voltages of other systems.)

Voltage (kV)	Distance above and below open supply conductors where clearances apply (ft)	Clearance between vertical conductor and pole center (in)
0 to 8.7	4	15
8.7 to 16	6	20
16 to 22	6	23
22 to 30	6	26
30 to 50	6	34

Table 239-1 Clearance of Vertical and Lateral Conductors M
(Circuit Phase-to-Phase Voltage)

Clearance of vertical and lateral conductors	0 to 8.7 kV (mm)	8.7 to 50 kV (mm)	Over 50 kV ④ (mm)
From surfaces of supports	①②75	75 plus 5 per kV over 8.7 kV	280 plus 5 per kV over 50 kV
From span, guy, and messenger wires	⑤150	150 plus 10 per kV over 8.7 kV ③	580 plus 10 per kV over 50 kV ③

① A neutral conductor meeting Rule 230E1 may be attached directly to the structure surface.

② For supply circuits of 0 to 750 V this clearance may be reduced to 25 mm.

③ Multiplier may be reduced to 6.5 mm/kV for anchor guys.

④ The additional clearance for voltages in excess of 50 kV specified in Table 239-1 shall be increased 3% for each 300 m in excess of 1000 m above mean sea level.

⑤ For cables meeting Rule 230C3 and operating at 0–750 V, this may be reduced to 50 mm.

Table 239-2 Clearance Between Open Vertical Conductors and Pole Center M

(Voltages are phase to ground for effectively grounded circuits and those other circuits where all ground faults are cleared by promptly de-energizing the faulted section, both initially and following subsequent breaker operations. See the definition section for voltages of other systems.)

Voltage (kV)	Distance above and below open supply conductors where clearances apply (m)	Clearance between vertical conductor and pole center (mm)
0 to 8.7	1.20	380
8.7 to 16	1.80	500
16 to 22	1.80	580
22 to 30	1.80	650
30 to 50	1.80	860

Section 24. Grades of Construction

240. General

A. The grades of construction are specified in this section on the basis of the required strengths for safety. Where two or more conditions define the grade of construction required, the grade used shall be the highest one required by any of the conditions.

B. For the purposes of this section, the voltage values for direct-current circuits shall be considered equivalent to the rms values for alternating-current circuits.

241. Application of Grades of Construction to Different Situations

A. Supply Cables

For the purposes of these rules, supply cables are classified by two types as follows:

Type 1—Supply cables conforming to Rules 230C1, 230C2, or 230C3 shall be installed in accordance with Rule 261I1.

Type 2—All other supply cables are required to have the same grade of construction as open-wire conductors of the same voltage.

B. Order of Grades

The relative order of grades for supply and communication conductors and supporting structures is B, C, and N, Grade B being the highest. Grade D is specified only for communication lines, and here it is higher than Grade N. Grade D cannot be directly compared with Grades B and C, but Rule 241C3b provides for conditions when such a combination of construction requirements exists.

C. At Crossings

Wires, conductors, or other cables of one line are considered to be at crossings when they cross over another line, whether or not on a common supporting structure, or when they cross over or overhang a railroad track or the traveled way of a limited access highway. Joint-use or colinear construction in itself is not considered to be at crossings.

1. Grade of Upper Line

Conductors and supporting structures of a line crossing

over another line shall have the grade of construction specified in Rules 241C3, 242, and 243.

2. Grade of Lower Line
Conductors and supporting structures of a line crossing under another line need only have the grades of construction which would be required if the line at the higher level were not there.

3. Multiple Crossings
   a. Where a line crosses in one span over two or more other lines, or where one line crosses over a span of a second line, which span in turn crosses a span of a third line, the grade of construction of the uppermost line shall be not less than the highest grade which would be required of either one of the lower lines when crossing the other lower line.
   b. Where communication conductors cross over supply conductors and railroad tracks in the same span, the grades of construction shall be in accordance with those listed in Table 241-1. It is recommended that the placing of communication conductors above supply conductors generally be avoided unless the supply conductors are trolley-contact conductors and their associated feeders.

D. Conflicts (see Section 2: *structure conflict*)
The grade of construction of the conflicting structure shall be as required by Rule 243A5.

Table 241-1
Grades of Construction for Communication Conductors Crossing Over Railroad Tracks and Supply Lines

When Crossing Over	Communication Conductor Grades
Railroad tracks and supply lines of 0 to 750 V to ground, or Type 1 supply cables of all voltages	D
Railroad tracks and supply lines exceeding 750 V to ground	B

242. Grades of Construction for Conductors

The grades of construction required for conductors are given in Tables 242-1 and 242-2. For the purpose of these tables certain classes of circuits are treated as follows:

A. Constant-Current Circuit Conductors

The grade of construction for conductors of a constant-current supply circuit involved with a communication circuit and not in Type 1 cable shall be based on either its current rating or on the open-circuit voltage rating of the transformer supplying such circuit, as set forth in Tables 242-1 and 242-2. When the constant current supply circuit is in Type 1 cable, the grade of construction shall be based on its nominal full-load voltage.

B. Railway Feeder and Trolley-Contact Circuit Conductors

Railway feeder and trolley contact circuit conductors shall be considered as supply conductors for the purpose of determining the required grade of construction.

C. Communication Circuit Conductors Used Exclusively in the Operation of Supply Lines

Communication circuit conductors used exclusively in the operation of supply lines shall have their grade of construction determined as follows:

1. By the requirements for ordinary communication circuits when conforming to Rule 224A3.
2. By the requirements for supply circuits when defined by Rule 224A4.

D. Fire-Alarm Circuit Conductors

Fire-alarm circuit conductors shall be considered as other communication circuit conductors except that they shall always meet grade D construction where the span length is from 0 to 150 ft (45 m), and grade C construction where the span length exceeds 150 ft (45 m).

E. Neutral Conductors of Supply Circuits

Supply-circuit neutral conductors, which are effectively grounded throughout their length and are not located above supply conductors of more than 750 volts to ground, shall have the same grade of construction as supply conductors of not more than 750 volts to ground, except that they need not meet any insulation requirements. Other neutral conductors shall have the same grade of construction as the phase conductors of the supply circuits with which they are associated.

F. Surge Protection Wires

Surge protection wires shall be of the same grade of construction as the supply conductors with which they are associated.

Table 242-1 Grades of Construction for Supply Conductors Alone, at Crossing, or on the Same Structures with Other Conductors

(The voltages listed in this table are phase to ground values for: effectively grounded ac circuits, two wire grounded circuits, or center grounded dc circuits; otherwise phase to phase values shall be used. The grade of construction for supply conductors, as indicated across the top of the table, must also meet the requirements for any lines at lower levels except when otherwise noted.)

Supply conductors at higher levels ① / Conductors, tracks and rights of way at lower levels	Constant-potential supply conductors										Constant current supply conductors		Communication conductors used exclusively in the operation of and run as supply lines
	0—0.75 kV		0.75—8.7 kV				Exceeding 8.7 kV						
	Urban	Rural	Urban		Rural		Urban		Rural				
	Open or Cable	Open or Cable	Open	Cable	Open	Cable	Open	Cable	Open	Cable	Open	Cable	Open or cable
Exclusive private rights-of-way	N	N	②N	N	N	N	②N	②N	N	N	B, C, or N; see Rule 242A		C or N; see Rule 242C
Common or public rights-of-way	N	N	C	N	N	N	③C	C	N	N			
Railroad tracks and limited access highways	B	B	B	B	B	B	B	B	B	B	B	B	B

Constant potential supply conductors 0 to 750 V Open or cable	N	N	C N	N N	③C C	④C N	B, C, or N; see Rule 242A	B, C, or N; see Rule 242C
750 V to 8.7 kV Open	⑤C	N	C C	N N	③C C	N N		
Cable	N	N	C N	N N	③C C	N N		
Exceeding 8.7 kV Open	⑤B	⑤C	B B	N N	③C C	N N		
Cable	⑤C	N	C N	N N	③C C	N N		
Constant current supply conductors: Open or cable	B, C, or N; see Rule 242A						B, C, or N; see Rule 242A	B, C, or N; see Rules 242A and 242C
Communication conductors: Open or cable, used exclusively in the operation of supply lines⑩	B, C, or N; see Rule 242C						B, C, or N; see Rules 242A and 242C	B, C, or N; see Rule 242C
Communication conductor: Urban or rural, open or cable⑥	N	N	⑦⑧B C	⑦⑧B C	⑧B C	⑧B C	⑧⑨B / C or N; see Rule 242A	B, C, or N; see Rule 242C

①The words "open" and "cable" appearing in the headings have the following meanings as applied to supply conductors: Cable means the Type 1 cables described in Rule 241A; open means open wire and Type 2 cables.

②Lines that can fall outside the exclusive private rights-of-way shall comply with the grades specified for lines not on exclusive private rights-of-way.

③Supply conductors shall meet the requirements of grade B construction if the supply circuits will not

(continued on next page)

Footnotes for Tables 242-1 *(Continued)*

be promptly de-energized, both initially and following subsequent breaker operations, in the event of a contact with lower supply conductors or other grounded objects.

④ Grade N construction may be used if crossing over supply services only.

⑤ If the wires are service drops, they may have grade N sizes and tensions as set forth in Table 263-2.

⑥ Grade N construction may be used where the communication conductors consist only of not more than one insulated twisted-pair or parallel-lay conductor, or where service drops only are involved.

⑦ Grade C construction may be used if the voltage does not exceed 5.0 kV phase to phase or 2.9 kV phase to ground.

⑧ The supply conductors need only meet the requirements of grade C construction if both of the following conditions are fulfilled:

(1) The supply voltage will be promptly removed from the communication plant by de-energization or other means, both initially and following subsequent circuit breaker operations in the event of a contact with the communication plant.

(2) The voltage and current impressed on the communication plant in the event of a contact with the supply conductors are not in excess of the safe operating limit of the communication protective devices.

⑨ Grade C construction may be used if the current cannot exceed 7.5 A or the open-circuit voltage of the transformer supplying the circuit does not exceed 2.9 kV.

⑩ Communication circuits located below supply conductors shall not affect the grade of construction of the supply circuits.

Table 242-2 Grades of Construction for Communication Conductors Alone, or in Upper Position of Crossing or on Joint Poles

(The voltages listed in this table are phase to ground values for: effectively grounded ac circuits, two wire grounded circuits, or center grounded dc circuits; otherwise phase to phase values shall be used. The grade of construction for supply conductors, as indicated across the top of the table, must also meet the requirements for any lines at lower levels except when otherwise noted.)

(Placing of communication conductors at higher levels at crossings, or on jointly used poles should generally be avoided, unless the supply conductors are trolley-contact conductors and their associated feeders.)

Conductors, tracks, and rights-of-way at lower levels	Communication conductors (communication conductors, rural or urban, open or cable, including communication conductors run as such, but used exclusively in the operation of supply lines.)
Exclusive private right-of-way	N
Common or public rights-of-way	N
Railroad tracks and limited access highways	D
Constant potential supply conductors①	
0 to 750 V	
Open or cable	N
750 V to 2.9 kV	
Open or cable	C
Exceeding 2.9 kV	
Open	B
Cable	C
Constant current supply conductors①	
0 to 7.5 A	
Open②	C
Exceeding 7.5 A	
Open②	③B
Communication conductors, open or cable, used exclusively in the operation of supply lines	④B, C, or N
Communication conductors, open or cable, urban or rural	N

①The words "open" and "cable" appearing in the headlines have the following meaning as applied to supply conductors: Cable means Type 1 cables as described in Rule 241A1; open means open wire and also Type 2 cables, as described in Rule 241A2.

②Where constant current circuits are in Type 1 cable, the grade of construction shall be based on the nominal full-load voltage.

③Grade C construction may be used if the open circuit voltage of the transformer supplying the circuit does not exceed 2.9 kV.

④See Rule 242C.

243. Grades of Construction for Line Supports

A. Structures

The grade of construction shall be that required for the highest grade of conductors supported except as modified by the following:

1. The grade of construction of jointly used structures, or structures used only by communication lines, need not be increased merely because the communication wires carried on such structures cross over trolley-contact conductors of 0 to 750 volts to ground.
2. Structures carrying grade C or D fire-alarm conductors, where alone, or where concerned only with other communication conductors, need meet only the requirements of grade N.
3. Structures carrying supply service drops of 0 to 750 volts to ground shall have at least the grade of construction required for supply line conductors of the same voltage.
4. Where the communication lines cross over supply conductors and a railroad in the same span and grade B is required by Rule 241C3b for the communication conductors, due to the presence of railroad tracks, the grade of the structures shall be D.
5. The grade of construction required for a conflicting structure (first circuit) shall be determined from the requirements of Rule 242 for crossings. The conflicting structure's conductors (first circuit) shall be assumed to cross the other circuit's conductors (second circuit) for the purposes of determining the grade of construction required for the conflicting structure.

   NOTE: The resulting structure grade requirement could result in a higher grade of construction for the structure than for the conductors carried thereon.

B. Crossarms and Support Arms

The grade of construction shall be that required for the highest grade of conductors carried by the arm concerned except as modified by the following:

1. The grade of construction of arms carrying only communication conductors need not be increased merely because the conductors cross over trolley-contact conductors of 0 to 750 volts to ground.

2. Arms carrying grade C or D fire-alarm conductors, where alone or where concerned with other communication conductors, need meet only the requirements for grade N.
3. Arms carrying supply service drops of 0 to 750 volts to ground shall have at least the grade of construction required for supply line conductors of the same voltage.
4. Where communication lines cross over supply conductors and a railroad in the same span and grade B is required by Rule 241C3b for the communication conductors due to the presence of railroad tracks, the grade of the arm shall be D.

C. Pins, Armless Construction Brackets, Insulators, and Conductor Fastenings

The grade of construction for pins and armless construction brackets, insulators, and conductor fastenings shall be that required for the conductor concerned except as modified by the following:

1. The grade of construction need not be increased merely because the supported conductors cross over trolley-contact conductors of 0 to 750 volts to ground.
2. Grade N construction is sufficient when only grade C or D fire-alarm conductors or other communication conductors are concerned.
3. Supply service drops of 0 to 750 volts to ground only require the same grade of construction as supply-line conductors of the same voltage.
4. When grade B construction is required by Rule 241C3b for the communication conductors due to the presence of railroad tracks, grade D construction shall be used when supporting communication lines which cross over supply conductors and a railroad in the same span.
5. When communication conductors are required to meet grade B or C, only the requirements for mechanical strength for these grades is required.
6. Insulators for use on open conductor supply lines shall meet the requirements of Section 27 for all grades of construction.

Section 25. Loading for Grades B, C, and D

250. General Loading Requirements and Maps

A. General

1. It is necessary to assume the loadings which may be expected to occur on a line because of wind and ice during all seasons of the year. These weather loadings shall be the values of loading resulting from the application of Rules 250B or 250C. Where both rules apply, the required loading shall be the one that, when combined with the appropriate overload capacity factors, has the greater effect on strength requirements.
2. Where construction or maintenance loads exceed those imposed by Rule 250A1, which may occur more frequently in light loading areas, the assumed loadings shall be increased accordingly.
3. It is recognized that loadings actually experienced in certain areas in each of the loading districts may be greater, or in some cases, may be less than those specified in these rules. In the absence of a detailed loading analysis, no reduction in the loadings specified therein shall be made without the approval of the administrative authority.

B. Combined Ice and Wind Loading

Three general degrees of loading due to weather conditions are recognized and are designated as heavy, medium, and light loading. Figure 250-1 shows the districts in the states in which these loadings are normally applicable.

NOTE: The localities are classified in the different loading districts according to the relative simultaneous prevalence of wind velocity and thickness of ice which accumulates on wires. Light loading is for places where little, if any, ice accumulates on wires.

Table 250-1 shows the radial thickness of ice and the wind pressures to be used in calculating loadings. Ice is assumed to weigh 57 pounds per cubic foot (913 kg/m^3).

C. Extreme Wind Loading

If any portion of a structure or its supported facilities exceeds 60 ft (18 m) above ground or water level, the applicable horizontal wind speed of Fig 250-2, as determined by the linear interpolation, shall be used to calculate horizontal wind pressures. These pressures shall be applied to the entire structure and supported facilities without ice loading. The following formulas shall be used to calculate wind pressures on cylindrical surfaces:

pressure in lbf/ft^2 = 0.00256 $(v_{mi/h})^2$

pressure in pascals = 0.613 $(v_{m/s})^2$

Table 250-2 lists the conversions of velocities to pressures for typical wind speeds as calculated by the formulas listed above.

If no portion of the structure or its supported facilities exceeds 60 ft (18 m) above ground or water level, the provisions of this rule are not required.

Fig 250-2 is a wind map of the contiguous United States and Alaska reproduced from ANSI A58.1-1982 [6]. For Hawaii and Puerto Rico, the basic wind speeds are 80 mi/h and 95 mi/h, respectively.

NOTE: Wind velocity usually increases with height; therefore, experience may show that the wind pressures specified herein need to be further increased.

Fig 250-1
General Loading Map of United States with Respect to Loading of Overhead Lines

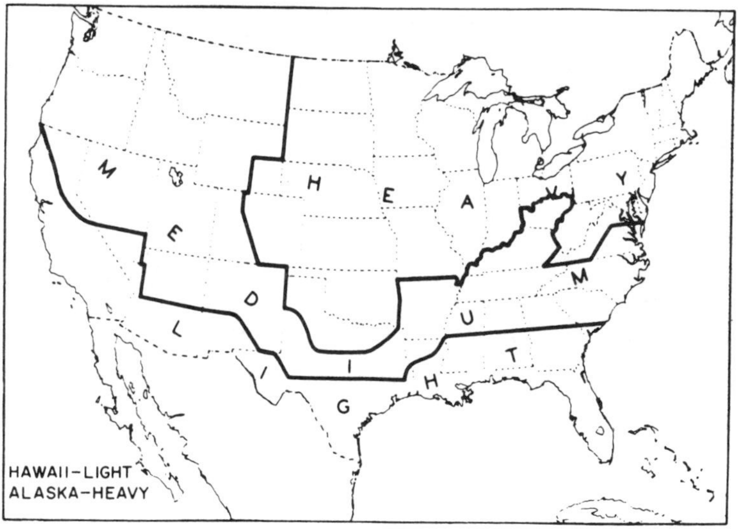

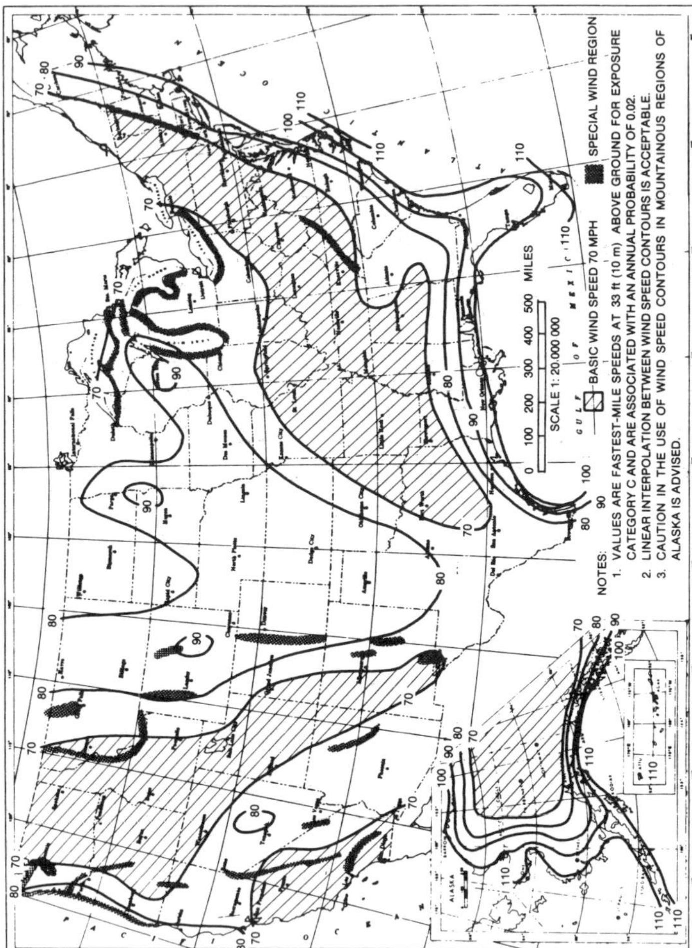

Fig 250-2 Basic Wind Speed (miles per hour)

Table 250-1 Ice, Wind and Temperature

	Loading Districts (For use with Rule 250B)			Extreme Wind Loading (For use with Rule 250C)
	Heavy	Medium	Light	
Radial thickness of ice (in)	0.50	0.25	0	0
(mm)	12.5	6.5	0	0
Horizontal wind pressure (lbs per sq ft)	4	4	9	See Fig 250-2
(Pa)	190	190	430	
Temperature (°F)	0	+15	+30	+60
(°C)	-20	-10	-1	+15

Table 250-2 Horizontal Wind Pressures on Cylindrical Surfaces

Wind Speed		Wind Pressure	
(mi/h)	(m/s)	(lb/ft^2)	(kPa)
70	31	13	0.60
80	36	16	0.78
90	40	21	1.00
100	45	26	1.23
110	49	31	1.48

251. Conductor Loading

A. General

Ice and wind loads shall be determined under Rule 250.

1. Where a cable is attached to a messenger, the specified loadings shall be applied to both cable and messenger.
2. In determining wind loadings on a bare stranded conductor or multiconductor cable, the assumed projected area shall be that of a smooth cylinder whose outside diameter is the same as that of the conductor or cable.

   NOTE: Experience has shown that as the size of multiconductor cable decreases, the actual projected area decreases, but the roughness factor increases and offsets the reduction in projected area.
3. In determining loadings on ice-covered bare stranded conductor or multiconductor cables, the coatings of ice shall be considered a hollow cylinder touching the outer strands of the bare stranded conductor or the outer circumference of the multiconductor cable. For bundled conductors, the coating of ice shall be considered as individual hollow cylinders around each subconductor.
4. It is recognized that the effects of conductor stranding or of non-circular cross section may result in wind and ice loadings more or less than those calculated by procedure stated in rule 251A2 and Rule 251A3. No reduction in these loadings is permitted unless tests or a qualified engineering study produces evidence that such a reduction is justifiable.

B. Loading Components

The components of loading and total loading shall be as follows.

1. Vertical Loading Component

   The vertical load on a conductor or messenger shall be its own weight plus the weight of conductors, spacers, or equipment which it supports, ice covered where determined under Rule 250.
2. Horizontal Loading Component

   The horizontal load shall be the horizontal wind pressure determined under Rule 250 applied at right angles to the direction of the line to the projected area of the conductor or messenger and conductors, spacers, or equipment which it supports, ice covered where determined under Rule 250.

3. Total Loading
The total load on a conductor or messenger shall be the resultant of components 1 and 2 above, calculated at the temperature specified in Table 251-1, to which resultant has been added the constant specified in Table 251-1. In all cases the conductor or messenger tension shall be computed from this total loading.

Table 251-1
Temperatures and Constants

	Loading districts (for use with Rule 250B)			Extreme wind loading (for use with Rule 250C)
	Heavy	Medium	Light	
Temperature (°F)	0	+15	+30	+60
(°C)	-20	-10	-1	+15
Constant to be added to the resultant (all conductors)				
in pounds per ft	0.30	0.20	0.05	0.0
in N/m	4.4	2.5	0.7	0.0

252. Loads Upon Line Supports

A. Assumed Vertical Loading
The vertical loads upon poles, towers, foundations, crossarms, pins, insulators, and conductor fastenings shall be their own weight plus the superimposed weight which they support, including all wires and cables, in accordance with Rules 251A and 251B1, together with the effect of any difference in elevation of supports. The radial thickness of ice shall be computed only upon wires, cables, and messengers, and not upon supports.

B. Assumed Transverse Loading
The total transverse loading upon poles, towers, foundations, crossarms, pins, insulators, and conductor fastenings shall include the following.

1. Transverse Loading from Conductors and Messengers
   The transverse loading from conductors and messengers shall be the horizontal loading determined under Rule 251.

   EXCEPTION: In medium- and heavy-loading districts, where supporting structures carry ten or more conductors on the same crossarm, not including cables supported by messengers, and where the horizontal pin spacing does not exceed 15 in (380 mm), the transverse wind loading may be calculated on two-thirds of the total number of such conductors if at least ten conductors are used in the calculations.

2. Structure Loading
   The transverse loading upon structures and equipment shall be computed by applying, at right angles to the direction of the line, the appropriate horizontal wind pressure determined under Rule 250. This pressure shall be applied upon the projected surfaces of the structures and equipment supported thereon, without ice covering. The following shape factors shall be applied.

   a. Cylindrical Structures and Components
      Wind loads on straight or tapered cylindrical structures or structures composed of numerous narrow relatively flat panels which combine to form a total cross section which is approximately circular or elliptical in shape shall be computed from the assumed unit wind pressure determined under Rule 250 applied to the projected area multiplied by a shape factor of 1.0.

   b. Flat Surfaced Structures and Components
      Wind loads on flat surfaced structures, having solid or enclosed flat sides and an overall cross section that is substantially square or rectangular, shall be computed from the assumed unit wind pressure determined under Rule 250 applied to the projected area of one face multiplied by a shape factor of 1.6 to allow for pressure on flat surfaces.

   c. Latticed Structures
      Wind loads on essentially square or rectangular latticed structures or components shall be computed from the assumed unit wind pressure determined under Rule 250 applied to the sum of the projected areas of the members of the front face multiplied by a shape factor of 3.2 to allow for wind pressure if

structural members are flat surfaced or 2.0 if structural surfaces are cylindrical. The total, however, need not exceed the load which would occur on a solid structure of the same outside dimension.

EXCEPTION: The shape factors listed under Rules 252B2a, 252B2b, and 252B2c may be reduced if wind tunnel tests or rational aerodynamic analysis produce evidence that such a reduction is justifiable.

3. At Angles

Where a change in direction of wires occurs, the loading upon the structure, including guys, shall be assumed to be a resultant load equal to the vector sum of the transverse wind load as derived above and the resultant load imposed by the wires due to their change in direction. In deriving these loadings, a wind direction shall be assumed which will give the maximum resultant load, proper reduction being made in loading to account for the reduced wind pressure on the wires resulting from the angularity of the application of the wind to the wires.

4. Span Lengths

The calculated transverse load shall be based upon the average of the actual lengths of the two spans adjacent to the structure concerned.

C. Assumed Longitudinal Loading

1. Change in Grade of Construction

The longitudinal loading upon supporting structures, including poles, towers, and guys at the ends of sections required to be of grade B construction, when located in lines of lower than grade B construction, shall be taken as an unbalanced pull in the direction of the higher grade section equal to the larger of the following values:

a. The pull of two-thirds, and in no case less than two of the conductors which have rated breaking strength of 3000 lb (13.3 kN) or less, such two-thirds of the conductors being selected so as to produce the maximum stress in the support.

b. The pull of one conductor when there are eight or less conductors (including overhead ground wires) having rated breaking strength of more than 3000 lb (13.3 kN), and the pull of two conductors when there are more than eight conductors, such conductors being selected so as to produce the maximum stress in the support.

2. Jointly Used Poles at Crossings Over Railroads, Communication Lines, or Limited Access Highways
Where a joint line crosses a railroad, a communication line, or a limited access highway, and grade B is required for the crossing span, the tension in the communication conductors of the joint line shall be considered as limited to one-half their rated breaking strength, provided they are smaller than Stl WG No 8, if of steel, or AWG No 6 if of copper.

3. Dead Ends
The longitudinal loading upon supporting structures at dead ends for line terminations shall be taken as an unbalanced pull equal to the tensions of all conductors and messengers (including overhead ground wires); except that with spans in each direction from the dead-end structure, the unbalanced pull shall be taken as the difference in tensions.

4. Unequal Spans and Unequal Vertical Loading
Where longitudinal loads can be created by the difference in tensions in the wires in adjacent spans caused by unequal vertical loading or unequal spans, the structures should be capable of supporting this unbalanced longitudinal loading.

5. Stringing Loads
Proper allowance should be made for longitudinal loads which may be produced on the structure by wire stringing operations.

6. Longitudinal Capability
It is recommended that structures having a longitudinal strength capability be provided at reasonable intervals along the line.

7. Communication Conductors on Unguyed Supports at Railroad Crossings and Limited Access Highways
The longitudinal loading shall be assumed equal to an unbalanced pull in the direction of the crossing of all open-wire conductors supported, the pull of each conductor being taken as 50% of its rated breaking

strength in the heavy loading district, 33 1/3% in the medium loading district, and 22 1/4% in the light loading district.

D. Simultaneous Application of Loads

Where a combination of vertical, transverse, or longitudinal loads may occur simultaneously, the structure shall be designed to withstand the simultaneous application of these loads.

NOTE: Under the extreme wind conditions of Rule 250C, an oblique wind may require greater structural strength than that computed under Rules 252B and 252C.

Section 26. Strength Requirements

260. Preliminary Assumptions (see also Section 20)

A. It is recognized that deformation, deflections, or displacement of parts of the structure will, in some cases, change the effects of the loads assumed. In the calculation of stresses, allowance may be made for such deformation, deflection, or displacement of supporting structures including poles, towers, guys, crossarms, pins, conductor fastenings, and insulators when the effects can be accurately evaluated. Such deformation, deflection, or displacement should be calculated using Rule 250 loads prior to application of the overload factors required by this section. For crossings or conflicts, the calculations shall be subject to mutual agreement.

B. It is recognized that newly developed materials may become available. It is further recognized that, while these materials are in the process of development, they must be tested and evaluated. Trial installations are permitted where qualified supervision is provided.

C. The overload capacity factors shown in the tables of this section apply for the combined ice and wind loading conditions specified in Rule 250B. For the extreme wind loading condition specified in Rule 250C, an overload capacity factor of not less than (a) 1.33 at installation and 1.0 at replacement for wood (including portion of wood poles below groundline) and reinforced concrete (not prestressed) structures, (b) 1.0 for all other structures and all foundations, and (c) 1.25 for other supported facilities shall be applied to the wind pressure corresponding to the wind speed determined from Figure 250-2.

261. Grades B and C Construction

A. Supporting Structure

The strength requirements for supporting structures may be met by the structures alone or with the aid of guys and/or braces.

1. Metal, Prestressed, and Reinforced Concrete Structures

   The structures shall be designed to withstand the loads in Rule 252 multiplied by the appropriate overload capacity factors given in Tables 261-1 or 261-2. (Where guys are used, see Rule 261C.)

   a. Minimum Strength

      All structures [including those below 60 ft (18 m)] shall be designed to withstand, without conductors, the extreme wind pressure in Rule 252 applied in any direction on the structure times an overload capacity factor of 1.0. A gust factor appropriate for the wind pressure and structure height should be considered.

   b. Strength at Angles in a Line

      At an angle in a line, the strength of the support shall be sufficient to withstand the total transverse loadings specified in Rule 252 multiplied by the appropriate overload capacity factor for transverse strength given in Tables 261-1 or 261-2.

2. Wood Structures

   Two methods for determining the capacity of wood structures are included herein. Either method meets the basic requirements for safety of wood structures.

   a. Method A

      Wood structures shall be of such material and dimensions as to meet the following requirements (where guys are used, see Rule 261C):

      (1) Designated Fiber Stress

         (a) Natural wood poles of various species meeting the requirements of ANSI O5.1-1987 [17], shall be considered as having the designated fiber stresses set forth in that standard.

(b) Appropriate adjustments in designated fiber stresses shall be made for sawn or laminated wood.

(2) Transverse and Vertical Strength
Wood structures shall be designed to withstand the transverse and vertical loads in Rule 252, multiplied by the appropriate overload capacity factor given in Table 261-3A, without exceeding the designated fiber stress.
EXCEPTION: When installed, naturally grown wood poles acting as single-based structures or unbraced multiple-pole structures, shall meet the requirements of Rules 261A2a(2) [261A2a(3)] without exceeding the designated fiber stress at the ground line for unguyed poles or at the points of attachment for guyed poles.

(3) Longitudinal and Dead-End Strength
Wood structures shall be designed to withstand the longitudinal and dead-end loadings in Rule 252 multiplied by the appropriate overload capacity factor in Table 261-3A without exceeding the designated fiber stress.
EXCEPTION 1: At a grade B crossing, in a straight section of line, wood structures complying with the transverse strength requirements of Rule 261A2a(2), without the use of transverse guys shall be considered as having the required longitudinal strength, providing the longitudinal strength is comparable to the transverse strength of the structure. This exception does not modify the requirements of this rule for dead ends.
EXCEPTION 2: At a grade B crossing of a supply line over a highway or a communication line where there is an angle in the supply line, wood structures shall be considered as having the required longitudinal strength if all of the following conditions are met:
(1) The angle is not over 20 degrees.
(2) The angle structure is guyed in the plane of the resultant of the conductor tensions. The tension in this guy under the loading in Rule 252 multiplied by an overload capacity factor of 2.0 shall not exceed the allowable guy value specified in Rule 261C.

(3) The angle structure has sufficient strength to withstand, without guys, the transverse loading of Rule 252, which would exist if there were no angle at that structure with an overload capacity factor of 4.0 when installed or 2.67 at replacement.

EXCEPTION 3: When installed, naturally grown wood poles acting as single based structures or unbraced multiple pole structures, shall meet the requirements of Rules 261A2a(2) [261A2a(3)] without exceeding the designated fiber stress at the ground line for unguyed poles or at the point of attachment for guyed poles.

(4) Strength at Angles in a Line

At an angle in the line, the wood structure shall be designed to withstand the total transverse loading in Rule 252 multiplied by the appropriate overload capacity factor given in Table 261-3A without exceeding the designated fiber stress.

(5) Strength of Guyed Poles

Guyed poles shall be designed as columns, resisting the vertical component of the tension in the guy plus any other vertical loads on such poles.

(6) Spliced and Stub-Reinforced Poles

The use of stub reinforcements or permanent splices at any section along the pole that develops the required strength of the pole is permitted, provided the remainder of the pole is in good condition and is of sufficient size to develop its required strength.

(7) Average Strength of Three Poles

A pole (single-base structure) not individually meeting the transverse strength requirements will be permitted when reinforced by a stronger pole on each side, if the average strength of the three poles meets the transverse strength requirements, and the weak pole has not less than 75% of the required strength. An extra pole inserted in a normal span for the purpose of supporting a service drop may be ignored.

EXCEPTION: This rule does not apply to crossings over railroads, communication lines, or limited access highways.

b. Method B

Wood structures shall be of such material and dimensions as to meet the following requirements (where guys are used, see Rule 261C):

(1) Permitted Stress Level

(a) Natural wood poles: The permitted stress level of natural wood poles of various species meeting the requirements of ANSI O5.1-1987, Specifications and Dimensions for Wood Poles [17], shall be determined by multiplying the designated fiber stresses set forth in that standard by the appropriate strength factor in Table 261-3B.

(b) Sawn or laminated wood structural members, crossarms, and braces: The permitted stress level of sawn or laminated wood structural members, crossarms, and braces shall be determined by multiplying the appropriate ultimate fiber stress of the material by the appropriate strength factor in Table 261-3B.

(2) Strength: Wood structures shall be designed to withstand the loads in Rule 252, multiplied by the appropriate overload capacity factors given in Table 261-3B, without exceeding the permitted stress level.

EXCEPTION 1: When installed, naturally grown wood poles acting as single-based structures or unbraced multi-pole structures, shall meet the requirements of Rule 261A2b(2) without exceeding the permitted stress level at the ground line for unguyed poles or at the points of attachment for guyed poles.

EXCEPTION 2: At a grade B crossing in a straight section of line, wood structures complying with the transverse strength requirements of Rule 261A2b(2) without the use of transverse guys shall be considered as having the required longitudinal strength, providing the longitudinal strength is comparable to the transverse strength of the structure. This exception does not modify the requirements of this rule for dead ends.

EXCEPTION 3: At a grade B crossing of a supply line over a highway or communication line where there is an angle in the supply line, wood structures shall be considered as having the required longitudinal strength if all of the following conditions are met:

(1) The angle is not over 20 degrees.

(2) The angle structure is guyed in the plane of the resultant of the conductor tensions. The tension in this guy under the loading in Rule 252 multiplied by an overload capacity factor of 2.0 shall not exceed the allowable guy value specified in Rule 261C.

(3) The angle structure neglecting the supporting effects of any guys has sufficient strength to withstand the transverse loadings of Rule 252, which would exist if there were no angle at that structure, multiplied by the appropriate overload capacity factor given in Table 261-3B without exceeding the permitted stress level.

(3) Strength of Guyed Poles:

Guyed poles shall be designed as columns, resisting the vertical component of the tension in the guy plus any other vertical loads on such poles.

(4) Spliced and Stub-Reinforced Poles:

The use of stub reinforcements or permanent splices at any section along the pole that develops the required strength of the pole is permitted, provided the remainder of the pole is in good condition and is of sufficient size to develop its required strength.

(5) Average Strength of Three Poles:

A pole (single-base structure) not individually meeting the transverse strength requirements will be permitted when reinforced by a stronger pole on each side, if the average strength of the three poles meets the transverse strength requirements, and the weak pole has not less than 75% of the required strength. An extra pole inserted in a normal span for the purpose of supporting a service drop may be ignored.

EXCEPTION: This rule does not apply to crossing over railroads, communication lines, or limited access highways.

3. Transverse-Strength Requirements for Structures Where Side Guying Is Required, But Can Only Be Installed at a Distance

   Grade B: In the case of structures where, because of very heavy or numerous conductors or relatively long spans, the transverse-strength requirements of this section cannot be met except by the use of side guys or special structures, and if it is physically impractical to employ side guys, the transverse-strength requirements may be met by side-guying the line at each side of, and as near as practical to, the crossing, or other transversely weak structure, and with a distance between such side-guyed structures of not over 800 ft (250 m), provided that:

   a. The side-guyed structures for each such section of 800 ft (250 m) or less shall be constructed to withstand the calculated transverse load due to wind on the supports and ice-covered conductors, on the entire section between side-guyed structures.
   b. The line between such side-guyed structures shall be substantially in a straight line and the average length of span between the side-guyed structures shall not exceed 150 ft (45 m).
   c. The entire section between the transversely strong structures shall comply with the highest grade of construction concerned in the given section, except as to the transverse strength of the intermediate poles or towers.

   Grade C: The above provisions do not apply to grade C.

4. Longitudinal-Strength Requirements for Sections of Higher Grade in Lines of a Lower Grade Construction

   a. Methods of Providing Longitudinal Strength

      Grade B: The longitudinal-strength requirements for sections of line of higher grade in lines of a lower grade (for assumed longitudinal loading, see Rule 252) may be met by placing supporting structures of the required longitudinal strength at either end of the higher grade section of the line.

      Where this is impractical, the supporting structures of the required longitudinal strength may be located one or more span lengths away from the section of higher grade, within 500 ft (150 m) on either side and with not more than 800 ft (250 m) between the longitudinally strong structures, provided such structures and the line between them meet the requirements as to transverse strength and stringing of conductors, of the highest grade occurring in the section, and provided that the line between the longitudinally strong structure is approximately straight or suitably guyed.

The requirements may also be met by distributing the head guys over two or more structures on either side of the crossing, such structures and the line between them complying with the requirements for the crossing as to transverse strength and as to conductors and their fastenings. Where it is impractical to provide the longitudinal strength, the longitudinal loads shall be reduced by increasing the conductor sags. This may require greater conductor separations. (See Rule 235B.)

Grade C: The above provisions do not apply to grade C.

b. Flexible Supports

Grade B: When supports of the section of higher grade are capable of considerable deflection in the direction of the line, as with wood or concrete poles, or some types of metal poles and towers, it may be necessary to increase the normal clearances specified in Section 23, or to provide head guys or special reinforcement to prevent such deflection.

Flexible metal structures may have to be head-guyed or otherwise reinforced to prevent reduction in the clearances required in Section 23.

Grade C: The above provision does not apply to grade C.

B. Strength of Foundations and Settings

The loadings in Rule 252 multiplied by the overload factors given in Table 261-4 shall be applied to the structure. Foundations and settings shall be designed or be determined by experience to withstand the reactions resulting from these applied loadings.

NOTE: Excessive movement of foundations may reduce structure strength or impair clearances.

C. Strength of Guys and Guy Anchors

The general requirements for guys and guy insulators are covered under Rules 264 and 279A, respectively. Guy anchors shall withstand the loads in Rule 252 multiplied by the overload factors given in Table 261-4.

1. Metal and Prestressed Concrete Structures

   Guys shall be considered as an integral part of the structure and shall withstand the loads in Rule 252, multiplied by the overload factors given in Table 261-2, without exceeding 90% of the rated breaking strength of the guy.

2. Wood and Reinforced Concrete Poles and Structures

   When guys are used to meet the strength requirements they shall be considered as taking the entire load in the direction in which they act, the structure acting as a strut only, except for those structures considered to possess sufficient rigidity so that the guy can be considered an integral part of the structure.

   a. Guys shall be of such material and dimension to withstand the loads in Rule 252, multiplied by the overload capacity factors given in Table 261-4 without exceeding 90% of the rated breaking strength of the guy.

   NOTE: Excessive movement of guys and guy anchors may reduce structure strength or impair clearances.

D. Crossarms

1. Wood Crossarms

   Two methods for determining the capacity of wood crossarms are included herein. Either method meets the basic requirements for safety of a wood crossarm.

   a. Strength, Method A

      Crossarms shall be designed to withstand the loads specified in Rule 252 multiplied by the appropriate overload capacity factors given in Table 261-3A without exceeding the designated fiber stress (or ultimate strength).

   b. Strength, Method B

      (1) Crossarms shall be designed to withstand the loads specified in Rule 252 multiplied by the

appropriate overload capacity factors given in Table 261-3B without exceeding their permitted stress level.

(2) The permitted stress level of solid sawn wood crossarms shall be determined by multiplying their ultimate fiber stress by the strength factor in Table 261-3B.

c. Material and Size

Wood crossarms of selected Southern pine or Douglas fir shall have a cross section of not less than those shown in Table 261-6. Crossarms of other suitable timber may be used provided they are of equivalent strength.

2. Concrete and Metal Crossarms

Crossarms shall be designed to withstand the loads specified in Rule 252 multiplied by the appropriate overload factors in Table 261-1 or 261-2.

3. Crossarms of Other Materials

Crossarms should meet the strength requirements of Rule 261D1a, Wood Crossarms.

4. Additional Requirements

a. Longitudinal Strength

(1) General

(a) Crossarms shall also be designed to withstand 700 lb (3.1 kN) applied at the outer conductor attachment points.

(b) At each end of a transversely weak section, as described in Rule 261A3, the longitudinal load shall be applied in the direction of the weak section.

(2) Methods of Meeting Rule 261D4a(1)

Grade B: Where conductor tensions are limited to a maximum of 2000 lb (9.0 kN) per conductor, double wood crossarms having cross sections specified in Table 261-6 and properly assembled will be considered as meeting the strength requirements specified in Rule 261D4a(1).

Grade C: This requirement is not applicable.

(3) Location

At crossings, crossarms should be attached to the face of the structure away from the crossing, unless special bracing or double crossarms are used.

b. Bracing

Crossarms shall be securely supported by bracing, if necessary, so as to suport safely all expected loads to which they may be subjected in use including line personnel working on them.

c. Double Crossarms or Brackets

Grade B: Where pin type construction is used, double crossarms or a support assembly of equivalent strength shall be used at each crossing structure, at ends of joint use of conflict sections, at dead ends and at corners where the angle of departure from a straight line exceeds 20 degrees. Under similar conditions, where a bracket supports a conductor operated at more than 750 volts to ground and there is no crossarm below, double brackets shall be used.

EXCEPTION: The above does not apply where communication cables or conductors cross below supply conductors and either are attached to the same pole, or where supply conductors are continuous and of uniform tension in the crossing span and each adjacent span. This exception does not apply to railroad crossings and limited access highways except by mutual agreement.

Grade C: The above requirement is not applicable.

E. Number 261E not used in this edition.

F. Strength of Pin Type or Similar Construction and Conductor Fastenings

1. Longitudinal Strength

a. General

(1) Pin type or similar construction and ties or other conductor fastenings shall withstand the applicable longitudinal loads given in Rule 252, or 700 lb (3.1 kN) applied at the pin, whichever is greater.

(2) At each end of a transversely weak section as described in Rule 261A3, the longitudinal load shall be applied in the direction of the weak section.

b. Method of Meeting Rules 261F1a

Grade B: Where conductor tensions are limited to 2000 lb (9.0 kN) and such conductors are supported on pin insulators, double wood pins and ties or their equivalent, will be considered to meet the requirements of Rule 261F1a.

Grade C: No requirement.

c. At Dead Ends and at Ends of Higher Grade Construction in Line of Lower Grade

Grade B: Pins and ties or other conductor fastenings connected to the structure at a dead end or at each end of the higher grade section shall be of sufficient strength to withstand at all times without exceeding their ultimate strength, an unbalanced pull due to the conductor loading specified in Rule 251.

Grade C: This requirement is not applicable except for dead ends.

d. At Ends of Transversely Weak Sections

Grade B: Pins and ties or other conductor fastenings connected to the structure at each end of the transversely weak section as described in Rule 261A3 shall be such as to withstand at all times without exceeding their ultimate strength, the unbalanced pull in the direction of the transversely weak section of the conductor supported, under the loading prescribed in Rule 251.

Grade C: No requirement.

2. Double Pins and Conductor Fastenings

Grade B: Where wood pins are used, double pins and conductor fastenings shall be used where double crossarms or brackets are required by Rule 261D4c.

EXCEPTION: The above does not apply where communication cables or conductors cross below supply conductors and either are attached to the same pole, or where supply conductors are continuous and of uniform tension in a crossing span and each adjacent span. This exception does not apply in the case of railroad crossings and limited access highway crossings except by mutual agreement.

Grade C: No requirement.

3. Single Supports Used in Lieu of Double Wood Pins

A single conductor support and its conductor fastening when used in lieu of double wood pins shall develop strength equivalent to double wood pins and their conductor fastenings as specified in Rule 261F1a.

G. Armless Construction

1. General

Open conductor armless construction is a type of open conductor supply line construction in which conductors are individually supported at the structure without the use of crossarms.

2. Insulating Material

Strength of insulating material shall meet the requirements of Section 27.

3. Other Components

Strengths of other components shall meet the appropriate requirements of Rules 260 and 261.

H. Open Supply Conductors

1. Sizes of Supply Conductors

Supply conductors shall have a rated breaking strength and an overall diameter of metallic conductor not less than that of medium-hard-drawn copper of the AWG size shown in Table 261-7 except that conductors made entirely of bare or galvanized iron or steel shall have an overall diameter not less than Stl WG of the gage sizes shown.

EXCEPTION 1: At railroad crossings, for stranded conductors, other than those in which a central core is entirely covered by the outside wires, any individual wire of such a stranded conductor containing steel shall be not less than 0.100 inch in diameter if copper or aluminum clad and not less than 0.115 inch in diameter if otherwise protected or if bare.

EXCEPTION 2: Service drops of 0 to 750 volts to ground may have the sizes set forth in Rule 263E.

2. Sags and Tensions

Conductor sags shall be such that, under the assumed loading of Rule 251 for the district concerned, the tensions of the conductor shall not be more than 60% of its rated breaking strength. Also the tension at 60 °F (15 °C), without external load, shall not exceed the following percentages of the conductor rated breaking strength:

Initial unloaded tension	35%
Final unloaded tension	25%

EXCEPTION: In the case of conductors having a cross section of a generally triangular shape, such as cables composed of three wires, the final unloaded tension at 60 °F (15 °C) shall not exceed 30% of the rated breaking strength of the conductor.

NOTE 1: The above limitations are based on the use of recognized methods for avoiding fatigue failures by minimizing chafing and stress concentration. If such practices are not followed, lower tensions should be employed.

NOTE 2: The factors listed above apply for the loading conditions of Rule 250B. For extreme wind loading conditions, See Rule 260C.

3. Splices, Taps, and Dead-End Fittings
   a. Splices should be avoided in crossings and adjacent spans. If it is impractical to avoid such splices, they shall have sufficient strength to withstand the maximum tension resulting from the loads in Rule 251 multiplied by an overload factor of 1.65.
   b. Taps should be avoided in crossing spans but if required shall be of a type which will not impair the strength of the conductors to which they are attached.
   c. Dead-end fittings, including the attachment hardware, shall have sufficient strength to withstand the maximum tension resulting from the loads in Rule 251 multiplied by an overload factor of 1.65.
4. Trolley-Contact Conductors

   In order to provide for wear, no trolley-contact conductor shall be installed of less size than AWG No 0, if of copper, or AWG No 4, if of silicon bronze.

I. Supply Cable Messengers

1. Messengers shall be stranded and shall not be stressed beyond 60% of their rated breaking strength under the loadings specified in Rule 251.

   NOTE 1: There are no strength requirements for cables supported by messengers.

   NOTE 2: Bonding and grounding requirements for Type 1 supply cables are in Section 21.

   NOTE 3: The factor in Rule 261I1 applies for the loading conditions of Rule 251, except when the extreme wind loading conditions, Rule 260C, apply.

J. Open-Wire Communication Conductors

Open-wire communication conductors in grade B or C construction shall have the sizes and sags given in Rules 261H1 and 261H2 for supply conductors of the same grade.

EXCEPTION: When open-wire communication conductors in spans of 150 ft (45 m) or less are above supply circuits of 5 kilovolts or less between conductors, grade C sizes and sags may be replaced by grade D sizes and sags, except that where the supply conductors are trolley-contact conductors of 0 to 750 volts to ground, WG No 12 Stl may be used for spans of 0 to 100 ft (0 to 30 m), and Stl WG No 9 may be used for spans of 125 to 150 ft (38 to 45 m).

K. Communication Cables

1. Communication Cables

   There are no strength requirements for such cables supported by messengers.

2. Messenger

   The messenger shall not be stressed beyond 60% of its rated breaking strength under the loadings specified in Rule 251.

L. Paired Communication Conductors

1. Paired Conductors Supported on Messenger

   a. Use of Messenger

      A messenger may be used for supporting paired conductors in any location, but is only required for paired conductors crossing over trolley-contact conductors of more than 7.5 kilovolts to ground.

   b. Sag of Messenger

      Messenger used for supporting paired conductors required to meet grade B construction because of crossing over trolley-contact conductors shall meet the sag requirements for grade D messengers.

   c. Size and Sag of Conductors

      There are no requirements for paired conductors when supported on messenger.

2. Paired Conductors Not Supported on Messenger

   a. Above Supply Lines

      Grade B: Sizes and sags shall be not less than those required by Rules 261H1 and 261H2 for supply conductors of similar grade.

      Grade C: Sizes and sags shall be not less than the following:

Spans 0 to 100 ft (0 to 30 m)—No sag requirements.

Each conductor shall have a rated breaking strength of not less than 170 lb (0.75 kN).

Spans 100 to 150 ft (30 to 45 m)—Sizes and sags shall be not less than required for grade D communication conductors.

Spans exceeding 150 ft (45 m)—sizes and sags shall be not less than required for grade C supply conductors. (See Rule 261H2).

b. Above Trolley-Contact Conductors

Grade B: Sizes and sags shall be not less than the following:

Spans 0 to 100 ft (0 to 30 m)—No size requirements. Sags shall be not less than for AWG No 8 hard-drawn copper. (See Rule 261H2.)

Spans exceeding 100 ft (30 m)—Each conductor shall have a rated breaking strength of not less than 170 lb (0.75 kN). Sags shall be not less than for AWG No 8 hard-drawn copper. (See Rule 261H2.)

Grade C: Sizes and sags shall be as follows:

Spans 0 to 100 ft (0 to 30 m)—No requirements.

Spans exceeding 100 ft (30 m)—No sag requirements.

Each conductor shall have a rated breaking strength of not less than 170 lb (0.75 kN).

Table 261-1 Overload Capacity Factors for Reinforced Concrete Structures (Not Prestressed)

	Overload capacity factors			
	Grade B		Grade C	
	When installed	At replacement ②	When installed	At replacement ②
Vertical loads ①	2.20	1.50	2.20	1.50
Transverse loads				
Wind	4.0	2.67	2.67	1.33
Wire tension at angles	2.0	1.33	1.33	1.00
Longitudinal loads				
In general	1.33	1.00	no requirement	
At dead ends	2.0	1.33	1.33	1.00

① Where vertical loading significantly reduces the loading on a structure member, a vertical overload factor of 1.0 should be used for the design of such member. Such members shall be designed for their worst case loading condition.

② When structure strength deteriorates to the level of the overload capacity factors required at replacement, the structure shall be replaced or rehabilitated. If a structure is replaced, it is required to meet the "when installed" overload capacity factors. Rehabilitated portions of structures shall have overload capacity factors in excess of those required "at replacement."

NOTE: The factors in this table apply for the loading conditions of Rule 250B. For extreme wind loading conditions, see Rule 260C.

Table 261-2 Overload Capacity Factors for Metal and Prestressed Concrete Structures

	Overload capacity factors	
	Grade B	Grade C
Vertical loads ①	1.50	1.50
Transverse loads		
Wind	2.50	2.20
Wire tension at angles	1.65	1.10
Longitudinal loads		
At Crossings		
In general	1.10	no requirement
At dead ends	1.65	1.10
Elsewhere		
In general	1.00	no requirement
At dead ends	1.65	1.10

① Where vertical loading significantly reduces the loading on a structure member, a vertical overload factor of 1.0 should be used for the design of such member. Such members shall be designed for their worst case loading condition.

NOTE: The factors in this table apply for the loading conditions of Rule 250B. For extreme wind loading conditions, see Rule 260C.

Table 261-3A Overload Capacity Factors for Wood Structures ①

	Grade B		Grade C	
	When installed	At replacement ②④	When installed	At replacement ③④
Vertical loads ③	2.20	1.50	2.20	1.50
Transverse (wind) loads				
At crossings	4.00	2.67	2.67	1.33
Elsewhere	4.00	2.67	2.00	1.33
Transverse (wire tension) loads				
At crossings	2.00	1.33	1.33	1.00
Elsewhere	2.00	1.33	1.33	1.00
Longitudinal loads				
In general	1.33	1.00	no requirement	
At dead ends	2.00	1.33	1.33	1.00

① Metal portions of a structure, except guys, may use the overload capacity factors for metal shown in Table 261-2.

② Where structures are built for temporary service, the overload capacity factors at replacement may be used provided that the designated fiber stress is not exceeded during the life of the structure.

③ Where vertical loading significantly reduces the loading on a structure member, a vertical overload factor of 1.0 should be used for the design of such member. Such members shall be designed for their worst case loading condition.

④ When structure strength deteriorates to the level of the overload capacity factors required at replacement, the structure shall be replaced or rehabilitated. If a structure is replaced, it is required to meet the "when installed" overload capacity factors. Rehabilitated portions of structures shall have overload capacity factors in excess of those required "at replacement."

NOTE: The factors in this table apply for the loading conditions of Rule 250B. For extreme wind loading conditions, see Rule 260C.

Table 261-3B Strength Factors and Overload Capacity Factors for Wood Structures When Installed ①②③④

	Grade B	Grade C
Strength factors	0.65	0.85
	Grade B	**Grade C**
Overload capacity factors:		
Vertical loads ⑤⑥	1.50	1.90
Transverse (wind) loads		
At crossings	2.50	2.20
Elsewhere	2.50	1.75
Transverse (wire tension) loads		
At crossings	1.65	1.30
Elsewhere	1.65	1.30
Longitudinal loads		
In general	1.0	no requirement
At dead ends	1.65	1.30

① Metal or concrete portions of a structure may meet the requirements of Rules 261A1 or 261C, as applicable.

② During the life of the structure, the capacity shall not be permitted to deteriorate to less than two-thirds of the capacity required by Rule 250B when installed.

③ The overload capacity factors in this table apply for loading conditions of Rule 250B. The overload capacity factors for extreme wind loading conditions of Rule 250C are stated in Rule 260C. The strength factors in this table shall be used for the loading conditions of Rule 250B and are not required for Rule 250C.

④ Where a structure is built for temporary service, its capacity throughout its life shall be not less than two-thirds of that required for a new structure when installed.

⑤ Where vertical loading significantly reduces the loading on a structure member, a vertical overload factor of 1.0 should be used for the design of such member. Such members shall be designed for their worst case loading condition.

⑥ It is intended that the resulting vertical capacity for Grades B and C be similar, as they are for Method A.

Table 261-4 Overload Capacity Factors for Guys, Guy Anchors, Foundations, and Settings

	Overload capacity factors	
	Grade B	Grade C
Vertical loads ①	1.50	1.50
Transverse loads		
Wind	2.50	2.20
Wire tension	1.65	1.10
Longitudinal loads		
In general	1.10	no requirement
At dead ends	1.65	1.10

① Where vertical loading significantly reduces the loading on a structure member, a vertical overload factor of 1.0 should be used for the design of such member. Such members shall be designed for their worst case loading condition.

NOTE: The factors in this table apply for the loading conditions of Rule 250B. For extreme wind loading conditions, see Rule 260C.

Table 261-5 not used in this edition.

Table 261-6 Dimensions of Crossarm Cross Section of Selected Southern Pine and Douglas Fir

		Grades of Construction		
		Grade B	Grade C	
Crossarm Length			Supply	Communication
4 ft (1.20 m)	in:	3 × 4	2¾ × 3¾	2¾ × 3¾
or less	mm:	75 × 100	70 × 95	70 × 95
8 ft (2.45 m)	in:	3¼ × 4¼	3 × 4	2¾ × 3¾
	mm:	82 × 108	75 × 100	70 × 95
10 ft (3.0 m)	in:	3¼ × 4¼	3 × 4	3 × 4
	mm:	82 × 108	75 × 100	75 × 100

Table 261-7 Conductor Sizes

Grade of Construction	Gage Size①
B	6
C	8

① For AWG no 6 and No 8 medium-hard-drawn copper wire, the nominal diameters are 0.1620 in and 0.1285 in, and the rated breaking strengths are 1010 lb (4.5 kN) and 643.9 lb (2.9 kN), respectively. For Stl WG, the nominal diameters are 0.192 in for No 6 and 0.162 for No 8.

262. Grade D Construction

A. Poles

Two methods for determining the capacity of wood poles are included herein. Either method meets the basic requirements for safety of a wood pole.

1. Method A
   a. Designated Fiber Stress
   Natural wood poles of various species meeting the requirements of ANSI O5.1-1987 [17] shall be considered as having the designated fiber stresses set forth in that standard.
   b. Strength of Unguyed Poles
   Unguyed poles shall withstand the vertical and transverse loads in Rules 252A and 252B, and the longitudinal loads in Rule 252C7, multiplied by the overload capacity factors given in Table 262-1A without exceeding the designated fiber stress.
   c. Strength of Guyed Poles
   Guyed poles shall be designed as columns, resisting the vertical component of the tension in the guy plus any other vertical loads on such poles.
   d. Spliced and Stub-Reinforced Poles
   The use of stub reinforcements or permanent splices at any section along the pole that develops the required strength of the pole is permitted, provided the remainder of the pole is in good condition and is of sufficient size to develop its required strength.
2. Method B
   a. Permitted Stress Level
   Natural wood poles: The permitted stress level of natural wood poles of various species meeting the requirements of ANSI O5.1-1987 Specifications and Dimensions for Wood Poles [17], shall be determined by multiplying the designated fiber stresses set forth in that standard by the appropriate strength factor in Table 262-1B.
   b. Strength of Unguyed Poles
   Unguyed poles shall be designed to withstand the vertical and transverse loads in Rules 252A and 252B, and the longitudinal loads in Rule 252C7,

multiplied by the overload capacity factors given in Table 262-1B without exceeding the permitted stress level.

c. Strength of Guyed Poles

Guyed poles shall be designed as columns, resisting the vertical component of the tension in the guy plus any other vertical loads on such poles.

d. Pole Replacement or Reinforcement

During the life of a structure, the capacity shall not be permitted to be less than two-thirds of that required by Rule 262A2.

B. Pole Settings

Foundations and settings for unguyed poles shall be such as to withstand the loads assumed in Rules 252A, 252B, and 252C.

C. Guys

1. General

The general requirements for guys are covered in Rules 264 and 279A.

2. Side Guys

a. Side guys or braces shall be installed on poles supporting the crossing span where required to withstand the loads specified in Rule 252.

EXCEPTION 1: Side guys are not required where the crossing poles have the transverse strength specified in Rule 262A1b or 262A2b, as applicable, without the reduction for conductor shielding otherwise allowed in Rule 252B1.

EXCEPTION 2: Where a line crossing a railroad or highway changes direction more than 10 degrees at either crossing support, the side guys within the angle may be omitted.

EXCEPTION 3: This rule does not apply to crossing poles under the special conditions set forth in Rule 262C5.

3. Longitudinal Guys

Longitudinal (head) guys shall be provided where required to meet the longitudinal strength requirements of Rule 252.

EXCEPTION: Longitudinal guys are not required where the crossing poles have the longitudinal strength specified in Rule 262A1b or 262A2b, as applicable, or for lines carrying only aerial cable. For lines carrying both open wire and aerial cable, head guying is required only for the number of open wires in excess of 10 if the cable is supported by a 6000 lb (26.7 kN) messenger, or for the number of open wires in excess of 20 if the cable is supported by a 10 000 lb (44.5 kN) or stronger messenger.

4. Strength of Guys
   a. Guys shall be of such material and dimensions to withstand the transverse and longitudinal loads in Rule 252, multiplied by the overload capacity factors given in Table 262-2, without exceeding 90% of their rated breaking strength.
   b. At an angle in the line, the guy shall be of such material and dimension to withstand the total transverse loads in Rule 252, multiplied by the overload capacity factors given in Table 262-3 without exceeding 90% of the rated breaking strength of the guy.
5. Where Guying Is Required But Cannot Be Installed on the Crossing Pole

   When the transverse-strength requirements cannot be met except by side guys and it is physically impractical to employ side guys, the transverse-strength requirements may be met by side-guying the line at each side of, and as near as is practical to, the crossing or other transversely weak structure, and with a distance between such side-guyed structures of not over 800 ft (250 m), provided that:
   a. The side-guyed structures for each such section of 800 ft (250 m) or less shall be constructed to withstand the calculated transverse load due to wind on the supports and ice covered conductors, on the entire section between the side-guyed structures.
   b. The line between such side-guyed structures shall be substantially in a straight line and the average length of span between the side-guyed structures shall not exceed 150 ft (45 m).
   c. The entire section between the transversely strong structures shall comply with the highest grade of construction concerned in the given section, except as to the transverse strength of the intermediate structures.

D. Crossarms

1. Material and Size

   Wood crossarms of southern pine or Douglas fir supporting the crossing span shall have a cross section not less than those shown in Table 262-4. Crossarms of other suitable timber or of other materials may be used provided they are of equivalent strength.

2. Double Crossarms

   Double crossarms or a support of equivalent strength shall be used at each crossing pole.

   EXCEPTION: Single dead-end type crossarms may be used where it is necessary to dead-end conductors of the crossing span, provided such crossarms and associated dead-end fastenings are of sufficient size and strength to withstand the maximum tension of the conductors under the loading specified in Rule 251 and provided further that the conductors are dead-ended on insulators so designed and installed that the conductor will not fall in the event of insulator breakage.

E. Brackets and Racks

Wood brackets may be used only in duplicate or if otherwise designed so as to afford two points of support for each conductor. Single metal brackets, racks, drive hooks or other fixtures may be used if designed and attached in such manner as to withstand the full dead-end pull of the wires supported.

F. Pins

1. Strength

   Insulator pins shall have sufficient strength to withstand all expected loads to which they may be subjected.

2. Size

   a. Wood pins

      Wood pins shall be sound and straight grained with a diameter of shank not less than 1 1/4 in (32 mm).

   b. Metal pins

      Steel or iron pins shall have diameters of shank not less than 1/2 in (12.5 mm).

G. Insulators

Each insulator shall be of such pattern, design, and material that when mounted it will withstand without injury and without being pulled off the pin, all expected loads to which they may be subjected.

H. Conductors

1. Size

Conductors of the crossing span, if of hard-drawn copper or galvanized steel, shall have sizes not less than given in the specifications a and b that follow. Conductors of material other than the above shall be of such size and so strung as to have a mechanical strength not less than that of the sizes of copper conductors given in specifications a and b that follow.

a. Ordinary Span Lengths

The sizes in Table 262-5 apply.

b. Long Spans

Long spans in excess of those specified in Table 262-5 are necessary, the size of conductors shall be increased so that the stress in the conductor will not exceed the limitations of Rule 262H3.

2. Paired Conductors Without Messengers

Paired wires without a supporting messenger shall be eliminated as far as practical but where used shall meet the following requirements.

a. Strength

Each conductor shall have a rated breaking strength of 170 lb (0.75 kN).

b. Limiting span lengths

Paired wires shall not be used without a supporting messenger in spans longer than 100 ft (30 m) in the heavy loading district, 125 ft (38 m) in the medium loading district, and 150 ft (45 m) in the light loading district.

3. Sags

Conductor sags shall be such that, under the assumed loading or Rule 251 for the district concerned, and assuming rigid structures for the purpose of calculations, the tension of the conductor shall not be more than 60% of its rated breaking strength. Also the final unloaded tensions at 60 °F (15 °C), shall not exceed 25% of the conductor rated breaking strength.

NOTE: The factors in Rule 262H3 apply for the loading conditions of Rule 250B. For extreme wind loading conditions see Rule 260C.

4. Splices and Taps

Splices shall, as far as practical, be avoided in the crossing and adjacent spans. If it is impractical to avoid such splices, they shall be of such type and so made as to have a strength substantially equal to that of the conductor in which they are placed.

Taps shall be avoided in the crossing span where practical, but if required shall be of a type which will not impair the strength of the conductors to which they are attached.

I. Messengers

1. Strength

Messengers shall be stranded material with a rated breaking strength of not less than 6000 lb (26.7 kN).

2. Sags and Tensions

Multiple-conductor cables and their messengers shall be so suspended that when they are subjected to the loading prescribed in Rule 251, the tension in the messenger shall not exceed 60% of its rated breaking strength.

NOTE: The factor in Rule 262I2 applies for loading conditions of Rule 251, except for extreme wind loading conditions where Rule 260C applies.

Table 262-1A Overload Capacity Factors for Unguyed Wood Poles ①

	Overload capacity factors
Vertical loads ③	
When installed	2.20
At replacement ②④	1.50
Transverse loads	
When installed	4.00
At replacement ② ④	2.67
Longitudinal loads	
When installed	1.33
At replacement ②④	1.00

① Metal portions of a structure, except guys, may use the overload capacity factors for metal shown in Table 261-2.

② Where structures are built for temporary service the overload factors at replacement may be used provided that the designated fiber stress is not exceeded during the life of the structure.

③ Where vertical loading significantly reduces the loading on a structure member, a vertical overload factor of 1.0 should be used for the design of such member. Such members shall be designed for their worst case loading condition.

④ When structure strength deteriorates to the level of the overload capacity factors required at replacement, the structure shall be replaced or rehabilitated. If a structure is replaced, it is required to meet the "when installed" overload capacity factors. Rehabilitated portions of structures shall have overload capacity factors in excess of those required "at replacement."

NOTE: The factors in this table apply for the loading conditions of Rule 250B. For extreme wind loading conditions, see Rule 260C.

Table 262-1B Strength Factor and Overload Capacity Factors for Unguyed Wood Poles When Installed ①②③④

Strength Factor	0.65
Overload capacity factors	
For vertical loads	1.5 ⑤
For transverse loads	2.5
For longitudinal loads	1.0

① Metal or concrete portions of a structure may meet the requirements of Rules 261A1 or 261C, as applicable.

② During the life of the structure, the capacity shall not be permitted to deteriorate to less than two-thirds of the capacity required for Rule 250B when installed.

③ The overload capacity factors in this table apply for loading conditions of Rule 250B. The overload capacity factors for extreme wind loading conditions of Rule 250C are stated in Rule 260C. The strength factor in this table shall be used for the loading conditions of Rule 250B and is not required for Rule 250C.

④ Where a structure is built for temporary service, its capacity throughout its life shall be not less than two-thirds of that required for a new structure when installed.

⑤ Where vertical loading significantly reduces the loading on a structure member, a vertical overload factor of 1.0 should be used for the design of such member. Such members shall be designed for their worst case loading condition.

Table 262-2 Overload Capacity Factors for Guys

	Overload capacity factors
Transverse strength	2.67
Longitudinal strength	
In general	1.0
At dead ends	1.5

NOTE: The factors in the table apply for the loading conditions of Rule 250B. For extreme wind loading conditions, see Rule 260C.

Table 262-3 Overload Capacity Factors for Guys at Angles in the Line

	Overload capacity factors Grade B
Transverse loads	
Wind	2.67
Wire tension	1.5

Table 262-4 Dimensions of Crossarm Cross Section of Southern Pine and Douglas Fir

Maximum number of wires to be carried	Nominal length			Cross section	
	(ft)	(in)	(m)	(in)	(mm) (mm)
2	1	4½	0.42	2 5/16 by 3 5/16	58 by 84
4	3	4½	1.00	2 5/16 by 3 5/16	58 by 84
6	6	0	1.80	2 3/4 by 3 3/4	70 by 95
10	8	6	2.60	2 3/4 by 3 3/4	70 by 95
10	10	0	3.0	3 by 4	75 by 100
① 12	10	0	3.0	3 1/4 by 4 1/4	82 by 108
② 16	10	0	3.0	3 1/4 by 4 1/4	82 by 108

① Where crossarms are bored for ½ in (12.5 mm) steel pins, 3 in by 4½ in (75 mm X 115 mm) crossarms may be used.

② Permitted in medium and light-loading districts only.

Table 262-5 Wire Sizes with Respect to Loading District and Span Length

	Spans			
	(ft)	(m)	(ft)	(m)
Heavy-loading district	0–125	0-38	126-150	38-45
Medium-loading district	0–150	0-45	151-175	45-53
Light-loading district	0–175	0-53	176-200	53-60
	Wire sizes			
Copper, hard drawn (AWG)	10		9	
Steel, galvanized (Stl WG)				
in general	10		8	
in rural districts of arid regions	12		10	
Aluminum or copper-clad steel (AWG)	10		9	

263. Grade N Construction

A. Poles

Poles used for lines for which neither grade B, C, or D is required shall be of such initial size and so guyed or braced, where necessary, as to withstand all expected loads to which they may be subjected, including line personnel working on them. Such poles and stubs on highways shall be located as far as is practical from the traveled portion of highways. The number of crossings over highways should be kept to a minimum. Such poles and stubs located within falling distance of the traveled way of highways, or so located that their failure would permit wires, cables, guys, or other equipment to fall into the traveled way of the highway, or would reduce the clearances specified in Table 232-1 over the highway, shall be periodically inspected and maintained in safe condition.

B. Guys

The general requirements for guys are covered in Rules 264 and 279A.

C. Crossarm Strength

Crossarms shall be securely supported by bracing, if necessary, to withstand all expected loads to which they may be subjected, including line personnel working on them.

NOTE: Double crossarms are generally used at crossings, unbalanced corners, and dead ends, in order to permit conductor fastenings at two insulators to prevent slipping, although single crossarms might provide sufficient strength. To secure extra strength, double crossarms are frequently used, and crossarm guys are sometimes used.

D. Supply-Line Conductors

1. Size

   Supply-line conductors shall be not smaller than the sizes listed in Table 263-1.

   RECOMMENDATION: It is recommended that these sizes for copper and steel be not used in spans longer than 150 ft (45 m) for the heavy-loading district, and 175 ft (53 m) for the medium- and light-loading districts.

E. Service Drops

1. Size of Open-Wire Service Drops

   a. Not over 750 volts. Service drops shall be as required by (1) or (2):

(1) Spans not exceeding 150 ft (45 m). Sizes shall not be smaller than those specified in Table 263-2.

(2) Spans exceeding 150 ft (45 m). Sizes shall not be smaller than required for grade C (Rule 261H1).

b. Exceeding 750 volts. Sizes of service drops of more than 750 volts shall not be less than required for supply-line conductors of the same voltage.

2. Tension of Open-Wire Service Drops

The tension of the service drop conductors shall not exceed the strength of the conductor attachment or its support under the expected loadings.

3. Cabled Service Drops

Service conductors may be grouped together in a cable, provided the following requirements are met:

a. Size

The size of each conductor shall not be less than required for drops of separate conductors (Rule 263E1).

b. Tension of Cabled Service Drops

The tension of the service drop conductors shall not exceed the strength of the conductor attachment or its support under the expected loadings.

F. Trolley-Contact Conductors

In order to provide for wear, no trolley-contact conductors shall be installed of less size than AWG No 0, if of copper, or AWG No 4, if of silicon bronze.

G. Communication Conductors

There are no specific requirements for grade N communication line conductors or service drops.

H. Street and Area Lighting Equipment

The lowering rope or chain for luminaires arranged to be lowered for examination or maintenance shall be of a material and strength designed to withstand climactic conditions and to sustain the luminaire safely.

Table 263-1 Sizes for Grade N Supply Line Conductors (AWG for Copper and Aluminum, Stl WG for Steel)

	Urban	Rural
Soft copper	6	8
Medium- or hard-drawn copper	8	8
Steel	9	9
	Spans 150 ft (45 m) or less	Spans exceeding 150 ft (45 m)
Stranded aluminum:		
EC	4	2
ACSR	6	4
ALLOY	4	4
ACAR	4	2

Table 263-2 Sizes of Service Drops of 750 V or Less

(Voltages of trolley-contact conductors are voltage to ground.)
AWG used for aluminum and copper wires; Stl WG used for steel wire.)

Situation	Copper wire: Soft drawn	Copper wire: Medium or hard drawn	Steel wire	EC aluminum wire②
Alone	10	12	12	4
Concerned with communication conductor	10	12	12	4
Over supply conductors of				
0 to 750 V	10	12	12	4
750 V to 8.7 kV①	8	10	12	4
Exceeding 8.7 kV①	6	8	9	4
Over trolley-contact conductors				
0 to 750 V ac or dc	8	10	12	4
Exceeeding 750 V ac or dc	6	8	9	4

① Installation of service drops of not more than 750 V above supply lines of more than 750 V should be avoided where practical.

② ACSR or high-strength aluminum alloy conductor size shall be not less than No 6.

264. Guying and Bracing

A. Where Used

When the loads to be imposed on supporting structures are greater than can be safely supported by the structures alone, additional strength shall be provided by the use of guys, braces, or other suitable construction. Such measures shall also be used where necessary to prevent undue increase of sags in adjacent spans as well as to provide sufficient strength for those supports on which the loads are considerably unbalanced, for example, at corners, angles, dead ends, large differences in span lengths, and changes of grade of construction.

B. Strength

The strength of the guy or brace shall meet the requirements of Section 26 for the applicable grade of construction. For guy wires conforming to ASTM Standards, the minimum breaking strength value therein defined shall be the rated breaking strength required in this code.

C. Point of Attachment

The guy or brace should be attached to the structure as near as is practical to the center of the conductor load to be sustained. However, on lines exceeding 8.7 kilovolts, the location of the guy or brace may be adjusted to minimize the reduction of the insulation offered by nonmetallic support arms and supporting structures.

D. Guy Fastenings

Guys having an ultimate strength of 2000 lb (9.0 kN) or more and subject to small radius bends should be stranded and should be protected by suitable guy thimbles or their equivalent. Cedar and other softwood poles around which any guy having an ultimate strength of 10 000 lb (44.5 kN) or more is wrapped should be protected by the use of suitable guy shims.

Where there is a tendency for the guy to slip off the shim, guy hooks or other suitable means of preventing this action should be used. Shims are not necessary in the case of supplementary guys, such as storm guys.

E. Guy Markers (Guy Guards)

The ground end of anchor guys, exposed to pedestrian traffic, shall be provided with a substantial and conspicuous marker.

NOTE: Visibility of markers can be improved by the use of color or color patterns which provide contrast with the surroundings.

F. Electrolysis

Where anchors and rods are subject to electrolysis, suitable measures should be taken to minimize corrosion from this source.

G. Anchor Rods

1. Anchor rods should be installed so as to be in line with the pull of the attached guy when under load.

   EXCEPTION: This is not required for anchor rods installed in rock or concrete.

2. The anchor and rod assembly shall have an ultimate strength not less than that required of the guy(s) by Rule 264B.

Section 27. Line Insulation

270. Application of Rule

These requirements apply only to open conductor supply lines.

NOTE 1: See Rule 243C6.

NOTE 2: See Rule 242E for insulation requirements for neutral conductors.

271. Material and Marking

Insulators for operation of supply circuits shall be made of wet process porcelain or other material which will provide equivalent or better electrical and mechanical performance. Insulators for use at or above 2.3 kilovolts between conductors shall be marked by the maker with his name or trademark and an identification mark or markings which will permit determination of the electrical and mechanical properties. The marking shall be applied so as not to reduce the electrical or mechanical strength of the insulator.

NOTE: The identifying marking can be either a catalog number, trade number, or other means so that properties of the unit can be determined either through catalogs or other literature.

272. Ratio of Flashover to Puncture Voltage

Insulators shall be designed so that the ratio of their rated low frequency dry flashover voltage to low frequency puncture voltage is in conformance with applicable American National Standards. When a standard does not exist, this ratio shall not exceed 75%.

The applicable American National Standards are:

ANSI C29.1-1982 [8].
ANSI C29.2-1983 [9].
ANSI C29.3-1986 [10].
ANSI C29.4-1984 [11].
ANSI C29.5-1984 [12].
ANSI C29.6-1984 [13].
ANSI C29.7-1983 [14].

EXCEPTION: Insulators specifically designed for use in areas of high atmospheric contamination may have a rated low frequency dry flashover voltage not more than 80% of their low frequency puncture voltage.

273. Insulation Level

The rated dry flashover voltage of the insulator or insulators, when tested in accordance with ANSI C29.1-1982 [8] shall be not less than that shown in Table 273-1, unless based on a qualified engineering study. Higher insulation levels than those shown in Table 273-1, or other effective means, shall be used where severe lightning, high atmospheric contamination, or other unfavorable conditions exist. Insulation levels for system voltages in excess of those shown shall be based on a qualified engineering study.

Table 273-1 Insulation Level Requirements

Nominal voltage (between phases) (kV)	Rated dry flashover voltage of insulators ① (kV)	Nominal voltage (between phases) (kV)	Rated dry flashover voltage of insulators ① (kV)
0.75	5	46	125
2.4	20	69	175
6.9	39	115	315
13.2	55	138	390
23.0	75	161	445
34.5	100	230	640

① Interpolate for intermediate values.

274. Factory Tests

Each insulator or insulating part thereof for use on circuits operating at or above 2.3 kilovolts between conductors shall be tested by the manufacturer in accordance with applicable American National Standards, or, where such standards do not exist, other good engineering practices to assure their performance.

The applicable American National Standards are listed in Rule 272.

275. Special Insulator Applications

A. Insulators for Constant-Current Circuits

Insulators for use on constant-current circuits shall be selected on the basis of the rated full load voltage of the supply transformer.

B. Insulators for Single-Phase Circuits Directly Connected to Three-Phase Circuits

Insulators used on single-phase circuits directly connected to three-phase circuits (without intervening isolating transformers) shall have an insulation level not less than that required for the three-phase circuit.

276. Protection Against Arcing and Other Damage

In installing and maintaining insulators and conductors, precautions shall be taken to prevent as far as is practical any damage which might render the conductors or insulators liable to fall. Precautions shall also be taken to prevent, as far as is practical, any arc from forming or prevent any arc which might be formed from injuring or burning any parts of the supporting structures, insulators, or conductors.

277. Mechanical Strength of Insulators

Insulators shall withstand all applicable loads specified in Section 25 except those of Rule 250C without exceeding the following percentages of their rated ultimate strength for the specified application.

Cantilever	40%
Compression	50%
Tension	50%

NOTE 1: The rated ultimate mechanical strength of suspension type insulators is considered to be the rated *combined mechanical and electrical strength.*
NOTE 2: See ANSI C29.1-1982 [8].

278. Aerial Cable Systems

A. Electrical Requirements
   1. Covered or insulated conductors not meeting the requirements of Rule 230C1, 230C2, or 230C3 shall be considered as bare conductors for all insulation requirements.
   2. The insulators or insulating supports shall meet the requirements of Rule 273.
   3. The systems shall be so designed and installed to minimize long term deterioration from electrical stress.

B. Mechanical Requirements
   1. Insulators other than spacers used to support aerial cable systems shall meet the requirements of Rule 277.
   2. Insulating spacers used in spacer cable systems shall withstand the loads specified in Section 25 (except those of Rule 250C) without exceeding 50% of their rated ultimate strength.

279. Guy and Span Insulators

A. Insulators
   1. Properties of Guy Insulators
      a. Material
         Insulators shall be made of wet process porcelain, wood, glass fiber reinforced plastic or other material of suitable mechanical and electrical properties.
      b. Electrical Strength
         The guy insulator shall have a rated dry flashover voltage at least double the nominal line voltage and a rated wet flashover voltage at least as high as the nominal line voltage between conductors of the guyed circuit. A guy insulator may consist of one or more units.
      c. Mechanical Strength
         The rated ultimate strength of the guy insulator shall be at least equal to the rated breaking strength of the guy in which it is installed.

2. Use of Guy Insulators
   a. Ungrounded guys attached to supporting structures carrying open supply conductors of more than 300 volts, or if exposed to such conductors, shall be insulated.

      NOTE: Guys grounded in accordance with Rule 215C2 need not be insulated.

      EXCEPTION: A guy insulator is not required if the guy is attached to a supporting structure on private right-of-way if all the supply circuits exceeding 300 volts meet the requirements of Rule 220B2.
   b. Insulators shall be installed as follows:
      (1) All insulators shall be located at least 8 ft (2.45 m) above the ground.
      (2) Where hazard would exist with one insulator, two or more guy insulators shall be placed so as to include, in so far as is practical, the exposed section of the guy between them.
      (3) Insulators shall be so placed that in case any guy sags down upon another, the insulators will not become ineffective.
3. Corrosion Protection

   An insulator in the guy strand used exclusively for the elimination of corrosion of metal in ground rods, anchors, anchor rods, or pipe in an effectively grounded system, shall not be classified as a guy insulator and shall not reduce the mechanical strength of the guy.

B. Span-Wire Insulators

1. Properties of Span-Wire Insulators
   a. Material

      Insulators shall be made of wet process porcelain, wood, fiberglass, or other material of suitable mechanical and electrical properties.
   b. Insulation Level

      The insulation level of span-wire insulators shall meet the requirements of Rule 274.

      A hanger insulator, where used to provide single insulation as permitted by Rule 279B2, shall meet the requirements of Rule 274.
   c. Mechanical Strength

      The rated ultimate strength of the span-wire

insulator shall be at least equal to the rated breaking strength of the span wire in which it is located.

2. Use of Span-Wire Insulators
   a. All span wires, including bracket span wires, shall have a suitable insulator (in addition to an insulated hanger if used) inserted between each point of support of the span wire and the luminaire or trolley-contact conductor supported.

      EXCEPTION 1: Single insulators, as provided by an insulated hanger, may be permitted when the span wire or bracket is supported on wood poles supporting only trolley, railway feeder, or communication conductors used in the operation of the railway concerned.

      EXCEPTION 2: Insulators are not required if the span wire is effectively grounded.

      EXCEPTION 3: This rule does not apply to insulated feeder taps used as span wires.
   b. In case insulated hangers are not used, the insulator shall be located so that in the event of a broken wire the energized part of the span wire cannot be reached from the ground.

Section number 28 not used in this edition.

PART 3. Safety Rules for the Installation and Maintenance of Underground Electric-Supply and Communication Lines

Section 30. Purpose, Scope, and Application of Rules

300. Purpose

The purpose of Part 3 of this code is the practical safeguarding of persons during the installation, operation, or maintenance of underground or buried supply and communication cables and associated equipment.

301. Scope

Part 3 of this code covers supply and communication cables and equipment in underground or buried systems. The rules cover the associated structural arrangements and the extension of such systems into buildings. It also covers the cables and equipment employed primarily for the utilization of electric power when such cables and equipment are used by the utility in the exercise of its function as a utility. They do not cover installations in electric supply stations.

Section 31. General Requirements Applying to Underground Lines

310. Referenced Sections

The Introduction (Section 1), Definitions (Section 2), List of Referenced Documents (Section 3) and Grounding Methods (Section 9) of this code shall apply to the requirements of Part 3.

311. Installation and Maintenance

A. Persons responsible for underground facilities shall be in a position to indicate the location of their facilities.

B. Reasonable advance notice should be given to owners or operators of other proximate facilities which may be adversely affected by new construction or changes in existing facilities.

312. Accessibility

All parts which must be examined or adjusted during operation shall be arranged so as to be readily accessible to authorized persons by the provision of adequate working spaces, working facilities, and clearances.

313. Inspection and Tests of Lines and Equipment

A. When in Service

1. Initial Compliance with Safety Rules
   Lines and equipment shall comply with these safety rules upon being placed in service.
2. Inspection
   Accessible lines and equipment shall be inspected by the responsible party at such intervals as experience has shown to be necessary.
3. Tests
   When considered necessary, lines and equipment shall be subjected to practical tests to determine required maintenance.
4. Record of Defects
   Any defects affecting compliance with this Code revealed by inspection, if not promptly corrected, shall be recorded; such record shall be maintained until the defects are corrected.
5. Remedying Defects
   Lines and equipment with recorded defects which would endanger life or property, shall be properly repaired, disconnected, or isolated.

B. When Out of Service

1. Lines Infrequently Used
   Lines and equipment infrequently used shall be inspected or tested as necessary before being placed into service.
2. Lines Temporarily Out of Service
   Lines and equipment temporarily out of service shall be maintained in a safe condition.

3. Lines Permanently Abandoned
Lines and equipment permanently abandoned shall be removed or maintained in a safe condition.

314. Grounding of Circuits and Equipment

A. Methods
The methods to be used for grounding of circuits and equipment are given in Section 9.

B. Conductive Parts to Be Grounded
Cable sheaths and shields (except conductor shields), equipment frames and cases (including pad-mounted devices), and conductive lighting poles shall be effectively grounded. Ducts and riser guards of conductive material which enclose electric supply lines shall be effectively grounded.
EXCEPTION: This rule does not apply to parts which are 8 ft (2.45 m) or more above readily accessible surfaces or are otherwise isolated or guarded.

C. Circuits

1. Neutrals
Primary neutrals, secondary and service neutrals, and common neutrals shall be effectively grounded as specified in Rule 314A.
EXCEPTION: Circuits designed for ground fault detection and impedance current limiting devices.
2. Other Conductors
Conductors, other than neutral conductors, which are intentionally grounded, shall be effectively grounded as specified in Rule 314A.
3. Surge Arresters
Surge arresters shall be effectively grounded as specified in Rule 314A.
4. Use of Earth as Part of Circuit
Supply circuits shall not be designed to use the earth normally as the sole conductor for any part of the circuit.

315. Communication Protective Requirements

A. Where Required
Where communications apparatus is handled by other than qualified persons, it shall be protected by one or more of the means listed in Rule 315B if such apparatus is permanently connected to lines subject to any of the following:

1. Lightning
2. Possible contact with supply conductors whose voltage exceeds 300 Volts
3. Transient rise in ground potential exceeding 300 volts
4. Steady-state induced voltage of a level which may cause personal injury

   NOTE: When communications cables will be in the vicinity of supply stations where large ground currents may flow, the effect of these currents on communications circuits should be evaluated.

B. Means of Protection

Where communications apparatus is required to be protected under Rule 315A, protective means adequate to withstand the voltage expected to be impressed shall be provided by insulation, protected where necessary by surge arresters. Severe conditions may require the use of additional devices such as auxiliary arresters, drainage coils, neutralizing transformers, or isolating devices.

316. Induced Voltage

Rules covering supply line influence and communication line susceptiveness have not been detailed in this code. Cooperative procedures are recommended to minimize steady state voltages induced from proximate facilities. Therefore, reasonable advance notice should be given to owners or operators of other known proximate facilities which may be adversely affected by new construction or changes in existing facilities.

Section 32. Underground Conduit Systems

NOTE: While it is often the practice to use *duct* and *conduit* interchangeably, *duct,* as used herein, is a single enclosed raceway for conductors or cable; *conduit* is a structure containing one or more ducts; and *conduit system* is the combination of conduit, conduits, manholes, handholes, and/or vaults joined to form an integrated whole.

320. Location

A. Routing

1. General

   a. Conduit systems should be subject to the least disturbance practical. Conduit systems extending parallel to other subsurface structures should not be located directly over or under other subsurface structures. If this is not practical, the rule on separation, as stated in Rule 230B, should be followed.

   b. Conduit alignment should be such that there are no protrusions which would be harmful to the cable.

   c. Where bends are required, the bending radius shall be sufficiently large to prevent damage to cable being installed in the conduit.

      RECOMMENDATION: The maximum change of direction in any plane between lengths of straight rigid conduit without the use of bends should be limited to 5 degrees.

2. Natural Hazards

   Routes through unstable soils such as mud, shifting soil, etc, or through highly corrosive soils, should be avoided. If construction is required in these soils, the conduit should be constructed in such a manner as to minimize movement and or corrosion or both.

3. Highways and Streets

   When conduit must be installed longitudinally under the roadway, it should be installed in the shoulder or, to the extent practical, within the limits of one lane of traffic.

4. Bridges and Tunnels

   The conduit system shall be located so as to minimize the possibility of damage by traffic. It should be located to provide safe access for inspection or maintenance of both the structure and the conduit system.

5. Crossing Railroad Tracks

   a. The top of the conduit system should be located not less than 36 in (900 mm) below the top of the rails of a street railway or 50 in (1.27 m) below the top of the rails of a railroad. Where unusual conditions exist or where proposed construction would interfere with existing installations, a greater depth than specified above may be required.

EXCEPTION: Where this is impractical, or for other reasons, this separation may be reduced by agreement between the parties concerned. In no case, however, shall the top of the conduit or any conduit protection extend higher than the bottom of the ballast section which is subject to working or cleaning.

b. At crossings under railroads, manholes, handholes, and vaults should not, where practical, be located in the roadbed.

6. Submarine Crossing

Submarine crossings should be routed, installed, or both so they will be protected from erosion by tidal action or currents. They should not be located where ships normally anchor.

B. Separation from Other Underground Installations

1. General

The separation between a conduit system and other underground structures paralleling it should be as large as necessary to permit maintenance of the system without damage to the paralleling structures. A conduit which crosses over another subsurface structure shall have a separation sufficient to prevent damage to either structure. These separations should be determined by the parties involved.

EXCEPTION: When conduit crosses a manhole, vault, or subway tunnel roof, it may be supported directly on the roof with the concurrence of all parties involved.

2. Separations Between Supply and Communications Conduit Systems

Conduit systems to be occupied by communications conductors shall be separated from conduit systems to be used for supply systems by

a. 3 in (75 mm) of concrete
b. 4 in (100 mm) of masonry
c. 12 in (300 mm) of well-tamped earth

EXCEPTION: Lesser separations may be used where the parties concur.

3. Sewers, Sanitary and Storm

a. If conditions require a conduit to be installed parallel to and directly over a sanitary or storm sewer, it may be done provided both parties are in agreement as to the method.

b. Where a conduit run crosses a sewer, it shall be designed to have suitable support on each side of the sewer to prevent transferring any direct load onto the sewer.

4. Water Lines

Conduit should be installed as far as is practical from a water main in order to protect it from being undermined if the main breaks. Conduit which crosses over a water main shall be designed to have suitable support on each side as required to prevent transferring any direct loads onto the main.

5. Fuel Lines

Conduit should have sufficient separation from fuel lines to permit the use of pipe maintenance equipment. Conduit and fuel lines shall not enter the same manhole.

6. Steam Lines

Conduit should be so installed as to prevent detrimental heat transfer between the steam and conduit systems.

321. Excavation and Backfill

A. Trench

The bottom of the trench should be undisturbed, tamped, or relatively smooth earth. Where the excavation is in rock, the conduit should be laid on a protective layer of clean tamped backfill.

B. Quality of Backfill

All backfill should be free of materials that may damage the conduit system.

RECOMMENDATION: Backfill within 6 in (150 mm) of the con-duit should be free of solid material greater than 4 in (100 mm) in maximum dimension or with sharp edges likely to damage it. The balance of backfill should be free of solid material greater than 8 in (200 mm) in maximum dimension. Backfill material should be adequately compacted.

322. Ducts and Joints

A. General

1. Duct material shall be corrosion-resistant and suitable for the intended environment.
2. Duct materials, the construction of the conduit, or both shall be designed so that a cable fault in one duct would not damage the conduit to such an extent that it would cause damage to cables in adjacent ducts.

3. The conduit system shall be designed to withstand external forces to which it may be subjected by the surface loadings set forth in rule 323A except that impact loading may be reduced one third for each foot (300 mm) of cover so no impact loading need be considered when cover is 3 ft (900 mm) or more.
4. The internal surface of the duct shall be free of sharp edges or burrs which could damage supply cable.

B. Installation

1. Restraint

   Conduit, including terminations and bends, should be suitably restrained by backfill, concrete envelope, anchors, or other means to maintain its design position under stress of installation procedures, cable pulling operations, and other conditions such as settling and hydraulic or frost uplift.
2. Joints

   Ducts shall be joined in a manner sufficient to prevent solid matter from entering the conduit line. Joints shall form a sufficiently continuous smooth interior surface between joining duct sections so that supply cable will not be damaged when pulled past the joint.
3. Externally Coated Pipe

   When conditions are such that externally coated pipe is required, the coating shall be corrosion resistant and should be inspected, tested, or both, to see that the coating is continuous and intact prior to backfill. Precautions shall be taken to prevent damage to the coating when backfilling.
4. Building Walls

   Conduit installed through a building wall shall have internal and external seals intended to prevent the entrance of gas into the building insofar as practical. The use of seals may be supplemented by gas-venting devices in order to minimize building up of positive gas pressures in the conduit.
5. Bridges

   a. Conduit installed in bridges shall include the capability to allow for expansion and contraction of the bridge.

b. Conduits passing through a bridge abutment should be installed so as to avoid or resist any shear due to soil settlement.

c. Conduit of conductive material installed on bridges shall be effectively grounded.

6. In Vicinity of Manholes

Conduit should be installed on compacted soil or otherwise supported when entering a manhole to prevent shear stress on the conduit at the point of manhole entrance.

323. Manholes, Handholes, and Vaults

A. Strength

Manholes, handholes, and vaults shall be designed to sustain all expected loads which may be imposed upon the structure. The horizontal design loads, vertical design loads, or both shall consist of dead load, live load, equipment load, impact, load due to water table, frost, and any other load expected to be imposed upon the structure, to occur adjacent to the structure, or both. The structure shall sustain the combination of vertical and lateral loading that produces the maximum shear and bending moments in the structure.

1. In roadway areas, the live load shall consist of the weight of a moving tractor-semitrailer truck illustrated in Fig 323-1. The vehicle wheel load shall be considered applied to an area as indicated in Fig 323-2. In the case of multilane pavements, the structure shall sustain the combination of loadings which result in vertical and lateral structure loadings which produce the maximum shear and bending moments in the structure.

   NOTE: Loads imposed by equipment used in road construction may exceed loads to which the completed road may be subjected.

2. In designing structures not subject to vehicular loading, the design live load shall be not less than 300 lb per sq ft (14.5 kPa).

3. Live loads shall be increased by 30% for impact.

4. When hydraulic, frost, or other uplift will be encountered, the structure shall either be of sufficient weight or so restrained as to withstand this force. The weight of equipment installed in the structure is not to be

considered as part of the structure weight.

5. Where pulling iron facilities are furnished, they should be installed to withstand twice the expected load to be applied to the pulling iron.

B. Dimensions

Manholes shall meet the following requirements: A clear working space sufficient for performing the necessary work shall be maintained. The horizontal dimensions of the clear working space shall be not less than 3 ft (900 mm). The vertical dimensions shall be not less than 6 ft (1.80 m) except in manholes where the opening is within 1 ft (300 mm), horizontally, of the adjacent interior side wall of the manhole.

EXCEPTION 1: Where one boundary of the working space is an unoccupied wall and the opposite boundary consists of cables only, the horizontal working space between these boundaries may be reduced to 30 in (750 mm).

EXCEPTION 2: In manholes containing only communications cables, equipment, or both, one horizontal dimension of the working space may be reduced to not less than 2 ft (600 mm) provided the other horizontal dimension is increased so that the sum of the two dimensions is at least 6 ft (1.80 m).

C. Manhole Access

1. Round access openings in a manhole containing supply cables shall be not less than 26 in (650 mm) in diameter. Round access openings in any manhole containing communication cables only, or manholes containing supply cables and having a fixed ladder which does not obstruct the opening, shall be not less than 24 in (600 mm) in diameter. Rectangular access openings should have dimensions not less than 26 in by 22 in (650 mm by 560 mm).
2. Openings shall be free of protrusions which will injure personnel or prevent quick egress.
3. Manhole openings shall be located so that safe access can be provided. When in the highway, they should be located outside of the paved roadway when practical. They should be located outside the area of street intersections and crosswalks whenever practical to reduce the traffic hazards to the workers working at these locations.
4. Personnel access openings should be located so that they are not directly over the cable or equipment. When these

openings interfere with curbs, etc, they can be located over the cable if one of the following is provided:

a. a conspicuous warning sign
b. a protective barrier over the cable
c. a fixed ladder

D. Covers

1. Manholes and handholes, when not being worked in, shall be securely closed by covers of sufficient weight or proper design so they cannot be easily removed without tools.
2. Covers should be suitably designed or restrained so that they cannot fall into manholes or protrude into manholes sufficiently far to contact cable or equipment.
3. Strength of covers and their supporting structure shall be at least sufficient to sustain the applicable loads of Rule 323A.

E. Vault and Utility Tunnel Access

1. Access openings shall be located so that safe access can be provided.
2. Personnel access openings in vaults should be located so that they are not directly over or open directly into equipment or cable. In vaults, other types of openings (not personnel access) may be located over equipment to facilitate work on, replacement or installation of equipment.
3. Where accessible to the public, access doors to utility tunnels and vaults shall be locked unless qualified persons are in attendance to prevent entry by unqualified persons.
4. Such doors shall be designed so that a person on the inside may exit when the door is locked from the outside.

EXCEPTION: This rule does not apply where the only means of locking is by padlock and the latching system is so arranged that the padlock can be closed on the latching system to prevent locking from the outside.

F. Ladder Requirements

Fixed ladders shall be corrosion resistant. Portable ladders shall be used in accordance with Rule 420J.

RECOMMENDATION: Ladders should conform to ANSI A14.1-1982 [2], ANSI A14.2-1982 [3], ANSI A14.3-1984 [4] or ANSI A14.5-1982 [5].

G. Drainage

Where drainage is into sewers, suitable traps or other means should be provided to prevent entrance of sewer gas into manholes, vaults, or tunnels.

H. Ventilation

Adequate ventilation to open air shall be provided for manholes, vaults, and tunnels, having an opening into enclosed areas used by the public. Where such enclosures house transformers, switches, regulators, etc, the ventilating system shall be cleaned at necessary intervals.

EXCEPTION: This does not apply to enclosed areas under water or in other locations where it is impractical to comply.

I. Mechanical Protection

Supply cables and equipment should be installed or guarded in such a manner as to avoid damage by objects falling or being pushed through the grating.

J. Identification

Manhole and handhole covers should have an identifying mark which will indicate ownership or type of utility.

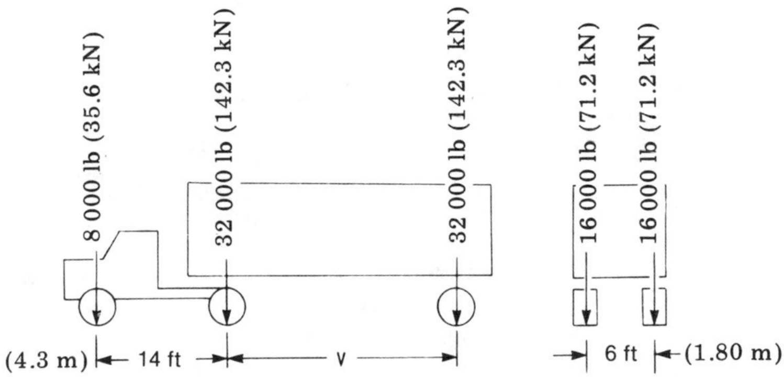

V = Variable spacing, 14 to 30 ft (4.3 to 9.0 m), inclusive. Spacing to be used is that which results in vertical and lateral structure loading which produces the maximum shear and bending moments in the structure.

Fig 323-1
Roadway Vehicle Load

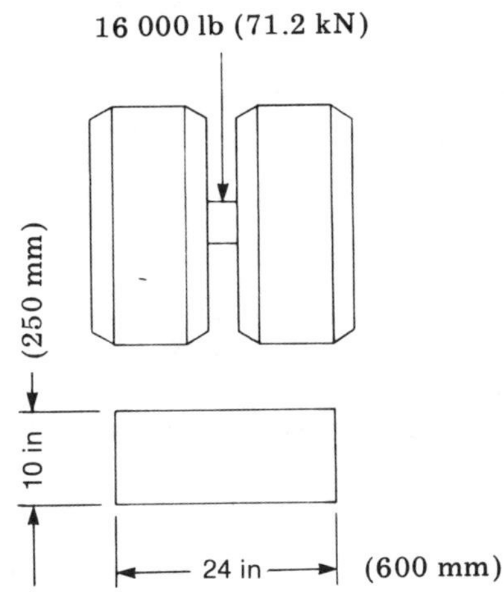

Fig 323-2
Wheel Load Area

Section 33. Supply Cable

330. General

RECOMMENDATION: Cable should be capable of withstanding tests applied in accordance with an applicable standard issued by a recognized organization such as the American National Standard Institute, the Association of Edison Illuminating Companies, the Insulated Cable Engineers Association, the National Electrical Manufacturers Association, or the American Society for Testing and Materials.

A. The design and construction of conductors, insulation, sheath, jacket, and shielding shall include consideration of mechanical, thermal, environmental, and electrical stresses which are expected during installation and operation.

B. Cable shall be designed and manufactured to retain specified dimensions and structural integrity during manufacture, reeling, storage, handling, and installation.

C. Cable shall be designed and constructed in such a manner that each component is protected from harmful effects of other components.

D. The conductor, insulation, and shielding shall be designed to withstand the effects of the expected magnitude and duration of fault current, except in the immediate vicinity of the fault.

331. Sheaths and Jackets

Sheaths, jackets, or both shall be provided when necessary to protect the insulation or shielding from moisture or other adverse environmental conditions.

332. Shielding

A. General

1. Conductor shielding should, and insulation shielding shall, be provided as specified by an applicable document issued by a nationally recognized cable standardization organization.

   NOTE: Typical cable standardization organizations include: the Association of Edison Illuminating Companies, the Insulated Cable Engineers Association, and the National Electrical Manufacturers Association.

   EXCEPTION: Shielding is not required for short jumpers which do not contact a grounded surface within enclosures or vaults, provided the jumpers are guarded or isolated.

2. Insulation shielding may be sectionalized provided that each section is effectively grounded.

B. Material

1. The shielding system may consist of semiconducting materials, nonmagnetic metal, or both. The shielding adjacent to the insulation shall be designed to remain in intimate contact with the insulation under all operating conditions.
2. Shielding material shall either be designed to resist excessive corrosion under the expected operating conditions or shall be protected.

333. Cable Accessories and Joints

A. Cable accessories and joints shall be designed to withstand the mechanical, thermal, environmental, and electrical stresses expected during operation.

B. Cable accessories and joints shall be designed and constructed in such a manner that each component of the cable and joint is protected from harmful effects of the other components.

C. Cable accessories and joints shall be designed and constructed to maintain the structural integrity of the cables to which they are applied and to withstand the magnitude and duration of the fault current expected during operation, except in the immediate vicinity of the fault.

D. For insulating joints, see Rule 332A2.

Section 34. Cable in Underground Structures

340. General

A. Section 33 shall apply to supply cable in underground structures.

B. On systems operating above 2 kV to ground, the design of the conductors or cables installed in nonmetallic conduit should consider the need for an effectively grounded shield, a sheath, or both.

341. Installation

A. General

1. Bending of the supply cable during handling, installation, and operation shall be controlled to avoid damage.

2. Pulling tensions and sidewall pressures on the supply cable should be limited to avoid damage.

   NOTE: Manufacturers' recommendations may be used as a guide.
3. Ducts should be cleaned of foreign material which could damage the supply cable during pulling operations.
4. Cable lubricants shall not be detrimental to cable or conduit systems.
5. On slopes or vertical runs, consideration should be given to restraining cables to prevent downhill movement.
6. Supply, control, and communication cables shall not be installed in the same duct unless the cables are maintained or operated by the same utility.

B. Cable in Manholes and Vaults

1. Supports
   a. Cable supports shall be designed to withstand both live and static loading and should be compatible with the environment.
   b. Supports shall be provided to maintain specified clearance between cables.
   c. Horizontal runs of supply cables shall be supported at least 3 in (75 mm) above the floor, or be suitably protected.

      EXCEPTION: This rule does not apply to grounding or bonding conductors.
   d. The installation should allow cable movement without destructive concentration of stresses. The cable should remain on supports during operation.

      NOTE: Special protection may be necessary at the duct entrance.
2. Clearance
   a. Adequate working space shall be provided in accordance with Rule 323B.
   b. Between supply and communication facilities (cable, equipment, or both):
      (1) Where cable, equipment, or both are to be installed in a joint-use manhole or vault, it shall be done only with the concurrence of all parties concerned.

(2) Supply and communication cables should be racked from separate walls. Crossings should be avoided.

(3) Where supply and communication cables must be racked from the same wall, the supply cables should be racked below the communication cables.

(4) Supply and communication facilities shall be installed to permit access to either without moving the other.

(5) Clearances shall be not less than those specified in Table 341-1.

3. Identification

a. General

(1) Cables shall be permanently identified by tags or otherwise at each manhole or other access opening of the conduit system.

EXCEPTION: This requirement does not apply where the position of a cable, in conjunction with diagrams or maps supplied to workers, gives sufficient identification.

(2) All identification shall be of a corrosion-resistant material suitable for the environment.

(3) All identification shall be of such quality and located so as to be readable with auxiliary lighting.

b. Joint-Use Manholes

(1) Where cables in a manhole are maintained or operated by different utilities or are of supply and communication usage, they shall be permanently marked as to company, type of use, or both.

Table 341-1 Clearance Between Supply and Communications Facilities in Joint-Use Manholes and Vaults

Phase-to-Phase Supply Voltage	Surface to Surface (in)	(mm)
0 to 15000	6	150
15 001 to 50000	9	230
50 001 to 120000	12	300
120 001 and above	24	600

EXCEPTION 1: These clearances do not apply to grounding conductors.
EXCEPTION 2: These clearances may be reduced by mutual agreement between the parties concerned when suitable barriers or guards are installed.

342. Grounding and Bonding

A. Insulation shielding of cable and joints shall be effectively grounded.

B. Cable sheaths or shields which are connected to ground at a manhole shall be bonded or connected to a common ground.

C. Bonding and grounding leads shall be of a corrosion-resistant material suitable for the environment or suitably protected.

343. Fireproofing

A. Although fireproofing is not a requirement, it may be provided in accordance with each utility's normal service reliability practice to provide protection from external fire.

344. Communication Cables Containing Special Supply Circuits

A. Special circuits operating at voltages in excess of 400 volts to ground and used for supplying power solely to communications equipment may be included in communications cables under the following conditions:

1. Such cables shall have a conductive sheath or shield which shall be effectively grounded and each such circuit shall be carried on conductors which are individually enclosed with an effectively grounded shield.
2. All circuits in such cables shall be owned or operated by one party and shall be maintained only by qualified personnel.
3. Supply circuits included in such cables shall be terminated at points accessible only to qualified employees.
4. Communications circuits brought out of such cables, if they do not terminate in a repeater station or terminal office, shall be protected or arranged so that in event of a failure within the cable, the voltage on the communications circuit will not exceed 400 V to ground.

5. Terminal apparatus for the power supply shall be so arranged that live parts are inaccessible when such supply circuits are energized.
6. Such cables shall be identified, and the identification shall meet the pertinent requirements of Rule 341B3.

   EXCEPTION: The requirements of Rule 344A do not apply to supply circuits of 550 volts or less which carry power not in excess of 3200 W.

Section 35. Direct Buried Cable

350. General

A. Section 33 shall apply to direct buried supply cable.

B. Cables operating above 600 V to ground shall have a continuous metallic shield, sheath or concentric neutral which is effectively grounded.

   EXCEPTION: At a splice or joint, the current path of the metallic shield, sheath, or neutral shall be made continuous but need not be concentric.

C. Cables of the same circuit operating below 600 V to ground and without an effectively grounded shield or sheath shall be placed in close proximity (no intentional separation) to each other.

D. Communications cables containing special circuits supplying power solely to communications equipment shall comply with the requirements of Rules 344A1 through 344A5.

E. Bonding should be provided between all above ground metallic power and communications apparatus (pedestals, terminals, apparatus cases, transformer cases, etc) that are separated by a distance of 6 ft (1.80 m) or less.

351. Location and Routing

A. General

1. Cables should be located so as to be subject to the least disturbance practical. Cables to be installed parallel to other subsurface structures should not be located directly over or under other subsurface structure, but if this is not practical, the rules on separations in Rule 352 should be followed.

2. Cables should be installed in as straight and direct a line as practical. Where bends are required, the bending radius shall be sufficiently large to prevent damage to the cable being installed.
3. Cable systems should be routed so as to allow safe access for construction, inspection, and maintenance.
4. The location of structures in the path of the projected cable route shall, as far as practical, be determined prior to trenching, plowing, or boring operation.

B. Natural Hazards

Routes through unstable soil such as mud, shifting soils, corrosive soils, or other natural hazards, should be avoided. If burying is required through areas with natural hazards, the cables shall be constructed and installed in such a manner as to protect them from damage. Such protective measures should be compatible with other installations in the area.

C. Other Conditions

1. Swimming Pools

   Supply cable should not be installed within 5 ft (1.50 m) of a swimming pool or its auxiliary equipment. If 5 ft (1.50 m) is not attainable, supplemental mechanical protection shall be provided.

2. Buildings and Other Structures

   Cable should not be installed directly under building or storage tank foundations. Where a cable must be installed under such a structure, the structure shall be suitably supported to prevent transfer of a harmful load onto the cable.

3. Railroad Tracks

   a. The installation of cable longitudinally under the ballast section for railroad tracks should be avoided. Where cable must be installed longitudinally under the ballast section of a railroad, it should be located at a depth of not less than 50 in (1.27 m) below the top of the rail.

      EXCEPTION: Where this is impractical, or for other reasons, this clearance may be reduced by agreement between the parties concerned.

      NOTE: Where unusual conditions exist or where proposed construction would interfere with existing installations, a greater depth than specified above would be required.

b. Where a cable crosses under railroad tracks, the same clearances indicated in Rule 320A5 shall apply.

4. Highways and Streets
   The installation of cable longitudinally under traveled surfaces of highways and streets should be avoided. When cable must be installed longitudinally under the roadway, it should be installed in the shoulder or, if this is not practical, within the limits of one lane of traffic to the extent practical.
5. Submarine Crossings
   Submarine crossings should be routed, installed, or both, so they will be protected from erosion by tidal action or currents. They should not be located where ships normally anchor.

352. Separations from Other Underground Structures (sewers, water lines, fuel lines, building foundations, steam lines, other supply or communication conductors not in random separation, etc)

A. Horizontal Separation
The horizontal separation between direct buried cable and other underground structures should be not less than 12 in (300 mm) to permit access to and maintenance of either facility without damage to the other. Installations with less than 12 in (300 mm) horizontal separation shall conform with requirements of Rule 352C, Rule 354, or both.

B. Crossings
1. Where a cable crosses under another underground structure, the structure shall be suitably supported to prevent transfer of a harmful load onto the cable system.
2. Where a cable crosses over another underground structure, the cable shall be suitably supported to prevent transfer of a harmful load onto the structure.
3. Adequate support may be provided by installing the facilities with sufficient vertical separation.
4. Adequate vertical separation shall be maintained to permit access to and maintenance of either facility without damage to the other. A vertical separation of 12 in (300 mm) is, in general, considered adequate but the parties involved may agree to a lesser separation.

C. Parallel Facilities

If conditions require a cable system to be installed with less than 12 in (300 mm) horizontal separation or directly over and parallel to another underground structure (or another underground structure installed directly over and parallel to a cable), it may be done providing all parties are in agreement as to the method. Adequate vertical separation shall be maintained to permit access to and maintenance of either facility without damage to the other.

D. Thermal Protection

Cable should be installed with sufficient separation from other underground structures, such as steam or cryogenic lines, to avoid thermal damage to the cable. Where it is not practical to provide adequate clearance, a suitable thermal barrier shall be placed between the two facilities.

353. Installation

A. Trenching

The bottom of the trench receiving direct buried cable should be relatively smooth undisturbed earth, well-tamped earth, or sand. When excavation is in rock or rocky soils, the cable should be laid on a protective layer of well-tamped backfill. Backfill within 4 in (100 mm) of the cable should be free of materials that may damage the cable. Backfill should be adequately compacted. Machine compaction should not be used within 6 in (150 mm) of the cable.

B. Plowing

1. Plowing in of cable in soil containing rock or other solid material should be done in such a manner that the solid material will not damage the cable, either during the plowing operation or afterward.
2. The design of cable plowing equipment and the plowing-in operation should be such that the cable will not be damaged by bending, side-wall pressure, or excessive cable tension.

C. Boring

Where a cable system is to be installed by boring and the soil and surface loading conditions are such that solid material in the region may damage the cable, the cable shall be adequately protected.

D. Depth of Burial

1. The distance between the top of a cable and the surface under which it is installed (depth of burial) shall be sufficient to protect the cable from damage imposed by expected surface usage.
2. Burial depths as indicated in Table 353-1 are considered adequate for supply cables or conductors, except as noted in a, b, or c following:
   a. In areas where frost conditions could damage cables, greater burial depths than indicated above may be desirable.
   b. Lesser depths than indicated above may be used where supplemental protection is provided. The supplemental protection should be sufficient to protect the cable from damage imposed by expected surface usage.
   c. Where the surface is not to final grade, under which a cable is to be installed, the cable should be placed so as to meet or exceed the requirements indicated above, both at the time of installation and subsequent thereto.

Table 353-1
Supply Cable or Conductor Burial Depth

Voltage (Phase-to-Phase)	Depth of Burial (in)	(mm)
0 to 600	24	600
601 to 50 000	30	750
50 001 and above	42	1070

EXCEPTION: Street light cables operating at not more than 150 V to ground may be buried at a depth not less than 18 in (450 mm).

354. Random Separation—Additional Requirements

These rules apply to cables or conductors when the radial separation between them will be less than 12 in (300 mm).

A. Supply Cables or Conductors

The cables or conductors of a supply circuit and those of another supply circuit may be buried together at the same depth with no deliberate separation between facilities, provided all parties involved are in agreement.

B. Communication Cables or Conductors

The cables or conductors of a communication circuit and those of another communication circuit may be buried together and at the same depth with no deliberate separation between facilities, provided all parties involved are in agreement.

C. Supply and Communication Cables or Conductors

Supply cables or conductors and communication cables or conductors may be buried together at the same depth with no deliberate separation between facilities, provided all parties involved are in agreement and the following requirements are met:

1. Voltage
   a. Grounded supply systems shall not be operated in excess of 22 000 volts to ground.
   b. Ungrounded supply systems shall not be operated in excess of 5300 volts phase to phase.
2. Grounded Conductor: Bare or with a Jacket of Radial Resistivity of 100 Meter Ohms or Less
   a. Where the phase conductor cable does not meet the requirements of Rule 354C3, a supply facility operating above 300 V to ground shall include a bare or semiconducting jacketed grounded conductor in continuous contact with the earth. This conductor, adequate for the expected magnitude and duration of the fault current which may be imposed, shall be one of the following:
      (1) A sheath, an insulation shield, or both
      (2) Multiple concentric conductors closely spaced circumferentially
      (3) A separate conductor in contact with the earth and in close proximity to the cable, where such cable or cables also have a grounded sheath or shield not necessarily in contact with the earth. The sheath, shield, or both, as well as the separate conductor, shall be adequate for the expected magnitude and duration of the fault currents which may be imposed.

      NOTE: This is applicable when a cable in nonmetallic duct is considered as a direct buried cable installation and random separation is desired.

EXCEPTION: Where buried cable passes through a short section of conduit such as under a roadway, the contact with earth of the grounded conductor can be omitted, provided the grounded conductor is continuous through the conduit.

b. The bare conductor or conductors in contact with the earth shall be of suitable corrosion resistant material. The conductor covered by a semiconducting jacket shall be compatible with the jacketing compound.

   NOTE: Experience has shown that for many geographic areas bare copper concentric neutral conductors experience severe corrosion.

c. The radial resistivity of the semiconducting jacket shall not be more than 100 meter ohms and shall remain essentially stable in service. The radial resistivity of the jacket material is that value calculated from measurements on a unit length of cable, of the resistance between the concentric neutral and a surrounding conducting medium. Radial resistivity is equal to the resistance of a unit length times the surface area of the jacket divided by the average thickness of the jacket over the neutral conductors. All dimensions are to be expressed in meters.

3. Grounded Conductor for Jacketed Cables with a Jacket Resistivity Exceeding 100 Meter Ohms

   Each phase conductor of a multiple grounded supply system operating above 300 volts to ground and having an overall insulating jacket with a radial resistivity exceeding 100 meter ohms shall have an effectively grounded copper concentric conductor meeting all of the following requirements:

   a. A conductance not less than one half that of the phase conductor
   b. Adequate for the expected magnitude and duration of fault current which may be imposed
   c. Grounded in accordance with Rule 314, except that the grounding interval required by Rule 96A3 shall be not less than eight in each mile of the random buried section, not including grounds at individual services.

4. Ungrounded Supply Systems

   Cables of an ungrounded supply system operating above

300 V shall be of effectively grounded concentric shield construction in continuous contact with the earth. Such cables shall be maintained in close proximity to each other.

D. Multiple Cable Systems

More than one cable system buried in random separation may be treated as one system when considering separation from other underground structures or facilities.

E. Protection

1. Supply circuits operating above 300 V to ground or 600 V between conductors shall be so constructed, operated, and maintained that when faulted, they shall be promptly de-energized initially or following subsequent protective device operation (phase-to-ground faults for grounded circuits, phase-to-phase faults for ungrounded circuits).
2. Ungrounded supply circuits operating above 300 V between conductors and in random separation with communication conductors shall be equipped with a ground fault indication system.
3. Communication-protective devices shall be adequate for the voltage and currents expected to be impressed on them in the event of contact with the supply conductors.
4. Adequate bonding shall be provided between the effectively grounded supply conductor or conductors and the communication cable shield or sheath at intervals which should not exceed 1000 ft (300 m).
5. In the vicinity of supply stations where large ground currents may flow, the effect of these currents on communication circuits should be evaluated before communication cables are placed in random separation with supply cables.

Section 36. Risers

360. General

A. Mechanical protection for supply conductors or cables shall be provided as required by Rule 239D of this code. This protection should extend at least 1 ft (300 mm) below ground level.

B. Supply conductors or cable should rise vertically from the cable trench with only such deviation as necessary to permit a reasonable cable-bending radius.

C. Exposed conductive pipes or guards containing supply conductors or cables shall be grounded in accordance with Rule 314.

361. Installation

A. The installation should be designed so that water does not stand in riser pipes above the frost line.

B. Conductors or cables shall be supported in a manner designed to prevent damage to conductors, cables, or terminals.

C. Where conductors or cables enter the riser pipe or elbow, they shall be installed in such a manner that shall minimize the possibility of damage due to relative movement of the cable and pipe.

362. Pole Risers—Additional Requirements

A. Risers should be located on the pole in the safest available position with respect to climbing space and possible exposure to traffic damage.

B. The number, size, and location of riser ducts or guards shall be limited to allow adequate access for climbing.

363. Pad-Mounted Installations

A. Supply conductors or cables rising from the trench to transformers, switchgear, or other equipment mounted on pads shall be so placed and arranged that they will not bear on the edges of holes through the pad nor the edges of bends or other duct work below the pad.

B. Cable entering pad-mounted equipment shall be maintained substantially at adequate depth for the voltage class until it becomes protected by being directly under the pad, unless other suitable mechanical protection is provided.

Section 37. Supply Cable Terminations

370. General

A. Cable terminations shall be designed and constructed to meet the requirements of Rule 333.

B. Riser terminations not located within a vault, pad-mounted equipment, or similar enclosure shall be installed in a manner designed to assure that clearance specified in Parts 1 and 2 of this code are maintained.

C. A cable termination shall be designed to prevent moisture penetration into the cable where such penetration is detrimental to the cable.

D. Where clearances between parts at different potentials are reduced below those adequate for the voltage and BIL (basic impulse insulation level), suitable insulating barriers or fully insulated terminals shall be provided to meet the required equivalent clearances.

371. Support at Terminations

A. Cable terminations shall be installed in a manner designed to maintain their installed position.

B. Where necessary, cable shall be supported or secured in a manner designed to prevent the transfer of damaging mechanical stresses to the termination, equipment, or structure.

372. Identification

Suitable circuit identification shall be provided for all terminations.

EXCEPTION: This requirement does not apply where the position of the termination, in conjunction with diagrams or maps supplied to workmen, gives sufficient identification.

373. Clearances in Enclosures or Vaults

A. Adequate electrical clearances of supply terminations shall be maintained, both between conductors and between conductors and ground, consistent with the type of terminator used.

B. Where exposed live parts are in an enclosure, clearances or insulating barriers adequate for the voltages and the design BIL shall be provided.

C. Where a termination is in a vault, uninsulated live parts are permissible provided they are guarded or isolated.

374. Grounding

A. All exposed conducting surfaces of the termination device, other than live parts and equipment to which it is attached, shall be effectively grounded, bonded, or both.

B. Conductive structures supporting cable terminations shall be effectively grounded.
EXCEPTION: Grounding, bonding, or both is not required where the above parts are isolated or guarded.

Section 38. Equipment

380. General

A. Equipment includes:
1. Buses, transformers, switches, etc, installed for the operation of the electric-supply system
2. Repeaters, loading coils, etc, installed for the operation of the communication system
3. Auxiliary equipment, such as sump pumps, convenience outlets, etc, installed incidental to the presence of the supply or communication systems

B. Where equipment is to be installed in a joint-use manhole, it shall be done with the concurrence of all parties concerned.

C. Supporting structures, including racks, hangers, or pads and their foundations shall be designed to sustain all loads and stresses expected to be imposed by the supported equipment including those stresses caused by its operation.

381. Design

A. The expected thermal, chemical, mechanical, and environmental conditions at the location shall be considered in the design of all equipment and mountings.

B. All equipment, including auxiliary devices, shall be designed to withstand the effects of normal, emergency, and fault conditions expected during operation.

C. Switches shall be provided with clear indication of contact position, and the handles or activating devices clearly marked to indicate operating directions.
RECOMMENDATION: The handles or control mechanism of all switches throughout the system should operate in a like direction to open and in a uniformly different direction to close in order to minimize errors.

D. Remotely controlled or automatic devices shall have provisions for local blocking to prevent operation if such operation may result in a hazard to the worker.

E. Enclosures containing fuses and interrupter contacts shall be designed to withstand the effects of normal, emergency, and fault conditions expected during operation.

F. When tools are to be used to connect or disconnect energized devices, space or barriers shall be designed to provide adequate clearance from ground or between phases.

G. Pad-Mounted Equipment

1. Pad-mounted equipment not located within a fenced or otherwise protected area shall have an enclosure that is either locked or secured against unauthorized entry.
2. Access to exposed live parts in excess of 600 volts shall require two separate conscious acts. The first shall be the opening of a door or barrier which is locked or otherwise secured against unauthorized entry. The second act shall be either the opening of a door or the removal of a barrier.

   RECOMMENDATION: A prominent Caution or other appropriate warning sign should be visible when the first door or barrier is opened or removed.

382. Location in Underground Structures

A. Equipment shall not obstruct personnel access openings in manholes or vaults nor shall it prevent easy egress by men working in the structures containing the equipment.

B. Equipment shall not be installed closer than 8 in (200 mm) to the back of fixed ladders and shall not interfere with the proper use of such ladders.

C. Equipment should be arranged in a manhole or vault to permit installation, operation, and maintenance of all items in such structures.

D. Switching devices which have provision for manual or electrical operation shall be operable from a safe position. This may be accomplished by use of portable auxiliary devices, temporarily attached.

E. Equipment should not interfere with drainage of the structure.

F. Equipment shall not interfere with the ability to ventilate any structure or enclosure.

383. Installation

A. Provisions for lifting, rolling to final position, and mounting shall be adequate for the weight of the device.

B. Live parts shall be guarded or isolated to prevent contact by persons in a normal position adjacent to the equipment.

C. Operating levers, inspection facilities, and test facilities shall be visible and readily accessible when equipment is in final location without moving permanent connections.

D. Live parts shall be isolated or protected from exposure to conducting liquids or other material expected to be present in the structure containing the equipment.

E. Operating controls of supply equipment, readily accessible to unauthorized personnel, shall be secured by bolts, locks, or seals.

384. Grounding

A. Cases and enclosures made of conductive material shall be effectively grounded or guarded.

B. Guards constructed of conductive material shall be effectively grounded.

385. Identification

Where transformers, regulators, or other similar equipment operate in multiple, tags, diagrams, or other suitable means shall be used to indicate that fact.

Section 39. Installation in Tunnels

390. General

A. The installation of supply and communication facilities in tunnels shall meet the applicable requirements contained elsewhere in Part 3 of this code as supplemented or modified by this section.

B. Where the space occupied by supply or communications facilities in a tunnel is accessible to other than qualified persons, or where supply conductors do not meet the requirements of Part 3 of this code for cable systems, the installation shall be in accordance with the applicable requirements of Part 2 of this code.

C. All parties concerned must be in agreement with the design of the structure and designs proposed for installations within it.

391. Environment

A. When the tunnel is accessible to the public or when workers must enter the structure to install, operate, or maintain the facilities in it, the design shall provide a controlled safe environment including where necessary, barriers, detectors, alarms, ventilation, pumps, and adequate safety devices for all facilities. Controlled safe environment shall include

1. Design to avoid poisonous or suffocation atmosphere
2. Design to protect persons from pressurized lines, fire, explosion, and high temperatures
3. Design to avoid unsafe conditions due to induced voltages
4. Design to prevent hazards due to flooding
5. Design to assure egress; two directions for egress shall be provided for all points in tunnels
6. Working space, in accordance with Rule 323B, the boundary of which shall be not less than 2 ft (600 mm) from a vehicular operating space or from exposed moving parts of machinery
7. Safeguards designed to protect workers from hazards due to the operation of vehicles or other machinery in tunnels
8. Unobstructed walkways for workers in tunnels

B. A condition of occupancy in multiple-use tunnels by supply and communications facilities shall be that the design and installation of all facilities is coordinated to provide a safe environment for the operation of supply facilities, communications facilities, or both. Safe environment for facilities shall include

1. Means to protect equipment from harmful effects of humidity or temperature
2. Means to protect equipment from harmful effects of liquids or gases
3. Coordinated design and operation of corrosion control systems.

PART 4. Rules for the Operation of Electric-Supply and Communications Lines and Equipment

Section 40. Purpose and Scope

400. Purpose

The purpose of Part 4 of this code is to provide practical work rules as one of the means of safeguarding employees and the public from injury.

401. Scope

Part 4 of this code covers work rules to be followed in the installation, operation, and maintenance of electric supply and communication systems.

402. Referenced Sections

The Introduction (Section 1), Definitions (Section 2), References (Section 3), and Grounding Methods (Section 9) of this code shall apply to the requirements of Part 4.

The American National Standards listed in Section 3 shall be used with Part 4 where applicable.

NOTE: After ANSI C2.4-1973 was originally approved June 30, 1972, OSHA (Occupational Safety and Health Administration) issued 29CFR1926 Subpart V applying to employee safety in construction. The revisions in the 1990 edition of the National Electrical Safety Code (ANSI C2-1990) did not address the differences. Footnotes have been provided in the text where OSHA requirements conflict with ANSI C2-1990 requirements.

Section 41. Supply and Communications Systems—Rules for Employers

410. General Requirements

A. General

1. The employer shall inform each employee working on or about communications equipment or electric-supply equipment and the associated lines, of the safety rules governing the employee's conduct while so engaged.

When deemed necessary, the employer shall provide a copy of such rules.

2. Employers shall utilize positive procedures to secure compliance with these rules. Cases may arise, however, where the strict enforcement of some particular rule could seriously impede the safe progress of the work at hand; in such cases the employee in charge of the work to be done should make such temporary modification of the rules as will accomplish the work without increasing the hazard.
3. If a difference of opinion arises with respect to the application of these rules, the decision of the employer or his authorized agent shall be final. This decision shall not result in any employee performing work in a manner which is unduly hazardous to the employee or to the employee's fellow workers.

B. Emergency Procedures and First Aid Rules

1. Employees shall be informed of procedures to be followed in case of emergencies and rules for first aid, including approved methods of resuscitation. Copies of such procedures and rules should be kept in conspicuous locations in vehicles and places where the number of employees and the nature of the work warrants.
2. Employees working on communications or electric-supply equipment or lines shall be regularly instructed in methods of first aid and emergency procedures, if their duties warrant such training.

C. Responsibility

1. A designated person shall be in charge of the operation of the equipment and lines and shall be responsible for their safe operation.
2. If more than one person is engaged in work on or about the same equipment or line, one person shall be designated as in charge of the work to be performed. Where there are separate work locations one person may be designated at each location.

411. Protective Methods and Devices

A. Methods

1. Access to rotating or energized equipment shall be restricted to authorized personnel.

2. Diagrams, showing plainly the arrangement and location of the electric-supply equipment and lines, shall be maintained on file and readily available to authorized personnel for that portion of the system for which they are responsible.
3. Employees shall be instructed as to the character of the equipment or lines and methods to be used before any work is undertaken thereon.
4. Employees should be instructed to take additional precautions to ensure their safety when conditions create unusual hazards.

B. Devices and Equipment

An adequate supply of protective devices and equipment, sufficient to enable employees to meet the requirements of the work to be undertaken, and first aid equipment and materials shall be available in readily accessible and, where practical, conspicuous places.

Protective devices and equipment shall conform to the applicable standards listed in Section 3.

NOTE: The following is a list of some common protective devices and equipment, the number and kinds of which will depend upon the requirements of each case:

1. Insulating wearing apparel such as rubber gloves, rubber sleeves, and headgear
2. Insulating shields, covers, mats, and platforms
3. Insulating tools for handling or testing energized equipment or lines
4. Protective goggles
5. *Person at work* tags, portable danger signs, traffic cones, and flashers
6. Body belts and safety straps
7. Fire extinguishing equipment designed for safe use on energized parts or plainly marked that they must not be so used
8. Protective grounding materials and devices
9. Portable lighting equipment
10. First aid equipment and materials

C. Inspection and Testing of Protective Devices

1. Protective devices and equipment shall be inspected or tested to ensure that they are in safe working condition.
2. Insulating gloves, sleeves, and blankets shall be inspected before use. Insulating gloves and sleeves shall be tested as frequently as their use requires.

3. Body belts, safety straps, and other personal equipment, whether furnished by employer or employee, shall be inspected to ensure that they are in safe working condition.

D. Warning Signs

Permanent warning signs shall be displayed in conspicuous places at all entrances to electric supply stations, substations, and other enclosed walk-in areas containing exposed current-carrying parts.

E. Identification of Supply Circuits

Means shall be provided so that identification of supply circuits can be determined before work is undertaken.

Section 42. General—Rules for Employees

420. Personal General Precautions

A. Rules and Emergency Methods

1. Employees shall carefully read and study the safety rules, and may be called upon at any time to show their knowledge of the rules.
2. Employees shall familiarize themselves with approved methods of first aid, rescue techniques, and fire extinguishment.

B. Qualifications of Employees

1. Employees whose duties require working on or in the vicinity of energized equipment or lines shall perform only those tasks for which they are trained, equipped, authorized, and so directed.

   Inexperienced employees shall: (a) work under the direction of an experienced and qualified person at the site, and (b) perform only directed tasks.
2. If an employee is in doubt as to the safe performance of any assigned work, the employee shall request instructions from the employee's supervisor or person in charge.
3. Employees who do not normally work on or in the vicinity of electric-supply lines and equipment but whose work brings them into these areas for certain tasks shall proceed with this work only when authorized by a qualified person.

C. Safeguarding Oneself and Others

1. Employees shall heed warning signs and signals and warn others who are in danger or in the vicinity of energized equipment or lines.
2. Employees shall report promptly to the proper authority any of the following:
   a. Line or equipment defects such as abnormally sagging wires, broken insulators, broken poles, or lamp supports
   b. Accidentally energized objects such as conduits, light fixtures, or guys
   c. Other defects that may cause a dangerous condition
3. Employees whose duties do not require them to approach or handle electric equipment and lines shall keep away from such equipment or lines and should avoid working in areas where objects and materials may be dropped by persons working overhead.
4. Employees who work on or in the vicinity of energized lines shall consider all of the effects of their actions, taking into account their own safety as well as the safety of other employees on the job site, or on some other part of the affected electric system, the property of others, and the public in general.
5. No employee shall approach or take any conductive object, without a suitable insulating handle, closer to any exposed energized part than allowed by Rule 431 (communication) or Rule 441 (supply) as applicable.
6. Employees should exercise care when extending metal ropes, tapes, or wires parallel to and in the proximity of energized high voltage lines because of induced voltages. When it is necessary to measure clearances from energized objects, only devices approved for the purpose shall be used.

D. Energized or Unknown Conditions

Employees shall consider electric supply equipment and lines to be energized, unless they are positively known to be de-energized. Before starting work, employees shall perform preliminary inspections or tests to determine existing conditions. Operating voltages of equipment and lines should be known before working on or in the vicinity of energized parts.

E. Ungrounded Metal Parts

Employees shall consider all ungrounded metal parts of equipment or devices, such as transformer cases and circuit breaker housings, to be energized at the highest voltage to which they are exposed, unless these parts are known by test to be free from such voltage.

F. Arcing Conditions

Employees should keep all parts of their bodies as far away as practical from switches, brushes, commutators, circuit breakers, or other parts at which arcing may occur during operation or handling.

G. Liquid Cell Batteries

1. Employees shall ascertain that battery areas are adequately ventilated before performing work.
2. Employees should avoid smoking, using open flames, or using tools which may produce sparks in the vicinity of liquid cell batteries.
3. Employees shall use eye and skin protection when handling an electrolyte.
4. Employees shall not handle energized parts of batteries unless necessary precautions are taken to avoid short circuits and electrical shocks.

H. Tools and Protective Equipment

Employees shall use the personal protective equipment, the protective devices, and the special tools provided for their work. Before starting work, these devices and tools shall be carefully inspected to make sure thay are in good condition.

I. Clothing

1. Employees shall wear clothing suitable for the assigned tasks and the work environment.
2. When working in the vicinity of energized lines or equipment, employees should avoid wearing exposed metal articles.

J. Ladders and Supports

1. Employees shall not support themselves, or any material or equipment, on any portion of a tree, pole structure, scaffold, ladder, walkway, or other elevated structure or aerial device, etc, without it first being determined, to the extent practical, that such support is adequately strong, in good condition, and properly secured in place.

2. Portable wood ladders intended for general use shall not be painted except with a clear nonconductive coating, nor shall they be longitudinally reinforced with metal.
3. Portable metal ladders intended for general use shall not be used when working on or in the vicinity of energized parts.
4. If portable ladders are made partially or entirely conductive for specialized work, necessary precautions shall be taken to ensure that their use will be restricted to the work for which they are intended.

K. Safety Straps

1. An employee working in an elevated position shall use a suitable safety strap or other approved means to prevent falling.
2. Safety straps or other similar devices shall be inspected before use by the employee to assure that they are in safe working condition.
3. Before employees trust their weight to safety straps or other devices, the employees shall determine that the snaps or fastenings are properly engaged and that the employees are secure in their body belts and safety straps.

L. Fire Extinguishers

In fighting fires or in the vicinity of exposed energized parts of electric supply systems, employees shall use fire extinguishers or materials which are suitable for the purpose. If this is not possible, all adjacent and affected equipment should first be de-energized.

M. Machines or Moving Parts

Employees working on normally moving parts of remotely controlled equipment shall be protected against accidental starting by proper tags installed on the starting devices, or by locking or blocking where practical. Employees shall, before starting any work, satisfy themselves that these protective devices have been installed. When working or in the vicinity of automatically or remotely operated equipment such as circuit breakers which may operate suddenly, employees shall avoid being in a position where they might be injured from such operation.

N. Fuses

When fuses must be installed or removed with one or both terminals energized, employees shall use special tools or gloves insulated for the voltage involved. When installing expulsion-type fuses, employees shall wear personal eye protection and take precautions to stand clear of the exhaust path of the fuse barrel.

O. Cable Reels

Cable reels shall be securely blocked so they cannot roll or rotate accidentally.

P. Street and Area Lighting

1. The lowering rope or chain, its supports, and fastenings shall be examined periodically.
2. A suitable device shall be provided by which each lamp on series lighting circuits of more than 300 volts may be safely disconnected from the circuit before the lamp is handled.

   EXCEPTION: This rule does not apply where the lamps are always worked on from suitable insulated platforms or aerial lift devices, or handled with suitable insulated tools, and treated as under full voltage of the circuit concerned.

421. General Operating Routines

A. Duties of a First Level Supervisor or Person in Charge

This individual shall:

1. Adopt such precautions as are within the individual's authority to prevent accidents.
2. See that the safety rules and operating procedures are observed by the employees under the direction of this individual.
3. Make all the necessary records and reports, as required.
4. Prevent unauthorized persons from approaching places where work is being done, as far as practical.
5. Prohibit the use of tools or devices unsuited to the work at hand, or which have not been tested or inspected as required.

B. Area Protection

1. Areas Accessible to Vehicular and Pedestrian Traffic
   a. Before engaging in work that may endanger the

public, warning signs or traffic control devices, or both, shall be placed conspicuously to alert approaching traffic. Where further protection is needed, suitable barrier guards shall be erected. Where the nature of work and traffic requires it, a person shall be stationed to warn traffic while the hazard exists.

b. When openings or obstructions in the street, sidewalk, walkways, or on private property are being worked on or left unattended during the day, danger signals, such as warning signs and flags, shall be effectively displayed. Under these same conditions at night, warning lights shall be prominently displayed and excavations shall be enclosed with protective barricades.

2. Areas Accessible to Employees Only

a. If the work exposes energized or moving parts that are normally protected, danger signs shall be displayed. Suitable barricades shall be erected to restrict other personnel from entering the area.

b. When working in one section where there is a multiplicity of such sections, such as one panel of a switchboard, one compartment of several, or one portion of a substation, employees shall mark the work area conspicuously and place barriers to prevent accidental contact with energized parts in that section or adjacent sections.

3. Locations with Crossed or Fallen Wires

An employee, finding crossed or fallen wires that are creating, or may create a hazard, shall remain on guard or adopt other adequate means to prevent accidents. The proper authority shall be notified. If the employee is qualified, and can observe the rules for safely handling energized parts by the use of insulating equipment, this employee may correct the condition.

C. Escort

Persons accompanying non-qualified employees or visitors or in the vicinity of electric equipment or lines shall be qualified to safeguard the people in their care, and see that the safety rules are observed.

422. Overhead Line Operating Procedures

Employees working on or with overhead lines shall observe the following rules in addition to applicable rules contained elsewhere in Sections 43 and 44.

A. Setting, Moving, or Removing Poles In or Near Energized Electric Supply Lines

1. When setting, moving, or removing poles in or in the vicinity of energized lines, precautions shall be taken to avoid direct contact of the pole with the energized conductors. Employees shall wear suitable insulating gloves or use other suitable means where voltages may exceed rating of gloves in handling poles where conductors energized at potentials above 750 V can be contacted. Employees performing such work shall not contact the pole with uninsulated parts of their bodies.
2. Contact with trucks, or other equipment, which are not bonded to an effective ground, being used to set, move, or remove poles in or in the vicinity of energized lines shall be avoided by employees standing on the ground or in contact with grounded objects unless employees are wearing suitable protective equipment.

B. Checking Structures Before Climbing

1. Before climbing poles, ladders, scaffolds, or other elevated structures, employees shall determine, to the extent practical, that the structures are capable of sustaining the additional or unbalanced stresses to which they will be subjected.
2. Where there are indications that poles and structures may be unsafe for climbing, they shall not be climbed until made safe by guying, bracing, or other means.

C. Installing and Removing Wires or Cables

1. Precautions shall be taken to prevent wires or cables being installed or removed from contacting energized wires or equipment. Wires or cables which are not bonded to an effective ground and are being installed or removed in the vicinity of energized conductors shall be considered as being energized.
2. Sag of wire or cables being installed or removed shall be controlled to prevent danger to pedestrian and vehicular traffic.

3. Before installing or removing wires or cables, the strains to which poles and structures will be subjected shall be considered and necessary action taken to prevent failure of supporting structures.
4. Employees should avoid contact with moving winch lines, especially in the vicinity of sheaves, blocks, and take-up drums.
5. Employees working on or in the vicinity of equipment or lines exposed to voltages higher than those guarded against by the safety appliances provided shall take steps to be assured that the equipment or lines on which the employee is working are free from dangerous leakage or induction or have been effectively grounded.

423. Underground Line Operating Procedures

Employees working on or with underground lines shall observe the following rules in addition to applicable rules contained elsewhere in Sections 43 and 44.

A. Guarding Manhole and Street Openings

When covers of manholes, handholes, or vaults are removed, the opening shall be promptly protected with a barrier, temporary cover, or other suitable guard.

B. Testing for Gas in Manholes and Unventilated Vaults

1. The atmosphere shall be tested for combustible or flammable gas(es) before entry.
2. Where combustible or flammable gas(es) are detected, the work area shall be ventilated and made safe before entry.
3. Unless forced continuous ventilation is provided, a test shall also be made for oxygen deficiency.
4. Provision shall be made for an adequate continuous supply of air.

   NOTE: The term adequate includes evaluation of both the quantity and quality of the air.

C. Flames

1. Employees shall not smoke in manholes.
2. Where open flames must be used in manholes or vaults, extra precautions shall be taken to ensure adequate ventilation.

3. Before using open flames in an excavation in areas where combustible gases or liquids may be present, such as in the vicinity of gasoline service stations, the atmosphere of the excavation shall be tested and found safe or cleared of the combustible gases or liquids.

D. Excavation

1. Cables and other buried utilities in the immediate vicinity shall be located, to the extent practical, prior to excavating.
2. Hand tools used for excavating in the vicinity of energized supply cables shall be equipped with handles made of nonconductive material.
3. Mechanized equipment should not be used to excavate in close proximity to cables and other buried utilities.
4. If a gas or fuel line is broken or damaged, employees shall:
   a. Leave the excavation open
   b. Extinguish flames which could ignite the escaping gas or fuel
   c. Notify the proper authority
   d. Keep the public away until the condition is under control

E. Identification

1. When underground facilities are exposed, they should be identified and shall be protected as necessary to avoid damage.
2. Where multiple cables exist in an excavation, cables other than the one being worked on shall be protected as necessary.
3. Before cutting into a cable or opening a splice, the cable should be identified and verified to be the proper cable.
4. When multiple cables exist in an excavation, the cable to be worked on shall be positively identified.

F. Operation of Power Driven Equipment

Employees should avoid being in manholes where power driven rodding equipment is in operation.

Section 43. Additional Rules for Communication Employees

430. General

Communication employees shall observe the following rules in addition to the rules contained in Section 42.

431. Approach to Energized Conductors or Parts

No employee shall approach, or take any conductive object, within the distances to any exposed energized part as listed in Table 431-1.

Table 431-1
Overhead Supply Lines and Equipment Approach Distances to Exposed Energized Parts

Voltage range (Phase-to-Phase, RMS)	Approach Distance (ft-in)	(m)
300 V and less	①	①
Over 300 V, not over 750 kV	1–0	0.30
Over 750 V, not over 2 kV	1–6	0.46
Over 2 kV, not over 15 kV	2–0	0.61
Over 15 kV, not over 37 kV	3–0	0.91
Over 37 kV, not over 87.5 kV	3–6	1.07
Over 87.5 kV, not over 121 kV	4–0	1.22
Over 121 kV, not over 140 kV	4–6	1.37

① Avoid contact.

432. Climbing Joint-Use Poles

Communication employees shall not climb poles where electric supply wires or equipment are hanging below their proper level. When working on jointly used poles or structures, employees shall not climb or work above the level of the lowest electric-supply conductor exclusive of vertical runs and street light wiring.

EXCEPTION: This rule does not apply where communications facilities are attached above electric-supply conductors if a rigid fixed barrier has been installed between the supply and communications facilities.

433. Attendant on Surface at Joint-Use Manhole

While personnel are in a joint-use manhole, an employee shall be available on the surface in the immediate vicinity to render assistance as may be required.

434. Sheath Continuity

Metallic or semiconductive sheath continuity shall be maintained by bonding across the opening, or by equivalent means, when working on buried cable or on cable in manholes.

Section 44. Additional Rules for Supply Employees

440. General

Supply employees shall observe the following rules in addition to the rules contained in Section 42.

441. Approach to Energized Conductors or Parts

An employee shall not approach, or willingly permit others to approach, any exposed ungrounded part normally energized except as permitted by Rule 441A, unless the supply equipment or lines are de-energized.

A. Clearance from Live Parts

1. Employees shall not approach, or take any conductive object without a suitable insulating handle, within the distances to the exposed parts which operate at the voltages listed in Table 441-1, or 441-3 unless at least one of the following is met: (1) the line or part is de-energized, or (2) the employee is insulated from the energized line or part, or (3) the energized line or part is insulated from the employee, or (4) the employee is insulated from all conducting surfaces other than the one upon which the employee is working.

   For voltages above 230 kV, ac, see Table 441-2. Gloves rated for the voltage involved shall be considered effective insulation of the employee from the energized part.

2. The additional clearances for voltages in excess of 50 kV specified in Rule 441A1 shall be increased 3 percent for each 1000 feet (300 meters) in excess of 3300 feet (1000 meters) above mean sea level.

B. If the conductor or part is to be de-energized, the requirements for authorization for switching, blocking, tagging, reporting, and grounding, as required in Section 44, shall be met.

Table 441-1 AC Live-Line Work Minimum Clearance

Nominal voltage in kilovolts phase to phase	Distance: Phase to employees (ft-in)	Distance: Phase to employees (m)	Distance: Phase to phase (ft-in)	Distance: Phase to phase (m)
1 to 34.5*	2-0	0.60	2-0	0.60
46	2-6	0.75	3-0	0.90
69 *	2-6	0.75	3-0	0.90
115 *	3-0	0.90	4-0	1.20
138	3-6	1.07	5-0	1.50
161 *	3-6	1.07	5-0	1.50
230 *	5-0	1.50	8-0	2.45
345 *	7-0	2.13	13-0	4.00
500	11-0	3.40	21-6	6.60
700	15-6	4.70	31-0	9.50

NOTE: These distances take into consideration the highest switching surge an employee will be exposed to on any system with air as the insulating medium and maximum operating voltages as stated in ANSI C84.1-1982 [15].

*OSHA 1926.955 specifies minimum clearances greater than ANSI C2.4-1973 as follows:

(kV)	Phase to employee (ft-in)	Phase to employee (m)	Phase to phase (ft-in)	Phase to phase (m)
15.1-35	2-4	0.71	2-4	0.71
69	3-0	0.90		
115	3-4	1.00		
161	3-8	1.12	5-6	1.70
230			8-4	2.50
345			13-4	4.10

Table 441-2 AC Live-Line Work Minimum Clearance with Surge Factor (See *NOTES* below) FT

Maximum switching surge factor	Distance in feet-inches								
	Phase to ground						Phase to phase (air, bare hand, or live-line tool)		
	Air/Bare-Hand			Live-Line tool					
	Nominal/maximum phase to phase kilovolts								
	345/362	500/550	700/765	345/362	500/550	700/765	345/362	500/550	700/765
1.5 or below	5-0/5-0	5-0/5-0	7-0/8-6	5-0/5-0	5-0/5-6	8-0/9-0	5-0/5-0	8-0/9-6	13-6/16-0
1.6	5-0/5-0	5-0/5-6	8-0/9-0	5-0/5-0	5-6/6-0	8-6/10-0	5-0/5-6	9-0/10-6	15-6/18-0
1.7	5-0/5-0	5-0/6-0	8-6/10-0	5-0/5-0	5-6/6-6	9-6/10-6	5-6/6-0	9-6/11-6	17-0/19-6
1.8	5-0/5-0	5-6/6-6	9-6/11-0	5-0/5-0	6-6/7-0	10-6/12-0	6-0/6-6	10-6/12-6	19-0/22-0
1.9	5-0/5-0	6-0/7-0	10-6/12-0	5-0/5-0	7-0/8-0	11-0/13-0	6-6/7-0	11-6/13-6	20-6/24-0
2.0	5-0/5-0	6-6/7-6	11-6/13-0	5-0/5-0	7-6/8-6	12-0/14-0	7-0/7-6	12-6/15-0	22-6/26-0
2.1	5-0/5-0	7-0/8-6	12-0/14-0	5-0/5-0	8-0/9-0	13-0/15-0	7-6/8-0	13-6/16-0	24-6/28-6
2.2	5-0/5-0	7-6/9-0	13-0/15-6	5-0/5-6	8-6/9-6	14-0/16-6	8-0/9-0	15-0/17-6	26-6/31-0
2.3	5-0/5-0	8-6/9-6		5-0/5-6	9-0/10-6		9-0/9-6	16-0/19-0	
2.4	5-0/5-6	9-0/10-0		5-6/6-0	9-6/11-0		9-6/10-0	17-6/20-0	
2.5	5-6/6-0	9-6/11-0		6-0/6-6	10-0/12-0		10-0/10-6	18-6/21-6	
2.6	5-6/6-0			6-0/6-6			10-6/11-6		
2.7	6-0/6-6			6-6/7-0			11-6/12-0		
2.8	6-6/7-0			7-0/7-6			12-0/13-0		

Footnotes for Table 441-2

NOTES: (1) The distances specified in this table may be applied *only* where the operating voltage and the switching surge factor are known and are supplied by the employer.

(2) On systems where the voltage is controlled to operate at a maximum value in between the nominal/maximum listed in the table, clearance distances may be determined by interpolation, rounding to the next higher 6 in.

(3) There is no provision in OSHA 1926 for clearance reduction through application of switching surge factors.

Table 441-2 AC Live-Line Work Minimum Clearance with Surge Factor (See *NOTES* below)

Maximum switching surge factor	Distance in meters								
	Phase to ground						Phase to phase (air, bare hand, or live-line tool)		
	Air/Bare-Hand			Live-Line tool					
	Nominal/maximum phase to phase kilovolts								
	345/362	500/550	700/765	345/362	500/550	700/765	345/362	500/550	700/765
1.5 or below	1.50/1.50	1.50/1.50	2.13/2.60	1.50/1.50	1.50/1.70	2.45/2.70	1.50/1.50	2.45/2.90	4.10/4.90
1.6	1.50/1.50	1.50/1.70	2.45/2.70	1.50/1.50	1.70/1.80	2.60/3.00	1.50/1.70	2.70/3.20	4.70/5.50
1.7	1.50/1.50	1.50/1.80	2.60/3.00	1.50/1.50	1.70/2.00	2.90/3.20	1.70/1.80	2.90/3.50	5.20/5.90
1.8	1.50/1.50	1.70/2.00	2.90/3.40	1.50/1.50	2.00/2.13	3.20/3.70	1.80/2.00	3.20/3.80	5.80/6.70
1.9	1.50/1.50	1.80/2.13	3.20/3.70	1.50/1.50	2.13/2.45	3.40/4.00	2.00/2.13	3.50/4.10	6.20/7.30
2.0	1.50/1.50	2.00/2.30	3.50/4.00	1.50/1.50	2.30/2.60	3.70/4.30	2.13/2.30	3.80/4.60	6.90/7.90
2.1	1.50/1.50	2.13/2.60	3.70/4.30	1.50/1.50	2.45/2.70	4.00/4.60	2.30/2.45	4.10/4.90	7.50/8.70
2.2	1.50/1.50	2.30/2.70	4.00/4.70	1.50/1.70	2.60/2.90	4.30/5.00	2.45/2.70	4.60/5.30	8.10/9.50
2.3	1.50/1.50	2.60/2.90		1.50/1.70	2.70/3.20		2.70/2.90	4.90/5.80	
2.4	1.50/1.70	2.70/3.00		1.70/1.80	2.90/3.40		2.90/3.00	5.30/6.10	
2.5	1.70/1.80	2.90/3.40		1.80/2.00	3.00/3.70		3.00/3.20	5.60/6.60	
2.6	1.70/1.80			1.80/2.00			3.20/3.50		
2.7	1.80/2.00			2.00/2.13			3.50/3.60		
2.8	2.00/2.13			2.13/2.30			3.60/4.00		

Footnotes for Table 441-2

NOTES: (1) The distances specified in this table may be applied *only* where the operating voltage and the switching surge factor are known and are supplied by the employer.

(2) On systems where the voltage is controlled to operate at a maximum value in between the nominal/maximum listed in the table, clearance distances may be determined by interpolation, rounding to the next higher 0.1 m.

(3) There is no provision in OSHA 1926 for clearance reduction through application of switching surge factors.

Table 441-3 DC Minimum Clearance for Live Parts (See *NOTES* below) **FT**

Transient overvoltage factor	Distance in feet-inches							
	Air				Live-line tool			
	Maximum conductor to ground kilovolts							
	250	400	500	750	250	400	500	750
1.5 or below	3-0	4-6	6-0	11-0	3-0	5-0	6-6	12-0
1.6	3-0	4-6	6-6	12-0	3-6	5-0	7-0	13-0
1.7	3-0	5-0	7-0	13-6	3-6	5-6	8-0	14-6
1.8 ④	3-6	5-6	7-6	15-0	3-6	6-0	8-6	16-0

NOTES: (1) The distances specified in this table may be applied *only* where the transient overvoltage factor is known and is supplied by the employer.

(2) On systems where the voltage is controlled to operate at a maximum value in between the maximums listed in the table, clearance distances may be determined by interpolation, rounding to the next higher 6 in.

(3) There is no provision in OSHA 1926 for clearance reduction through application of transient overvoltage factors.

④ These distances are based on the highest transient overvoltage an employee will be exposed to on any system with live-line tools as the insulating material. If the transient overvoltage factor is not known, the values for this factor shall be used.

Table 441-3 DC Minimum Clearance for Live Parts (See *NOTES* below) **M**

Transient overvoltage factor	Distance in meters							
	Air				Live-line tool			
	Maximum conductor to ground kilovolts							
	250	400	500	750	250	400	500	750
1.5 or below	0.90	1.40	1.80	3.40	0.90	1.50	2.00	3.70
1.6	0.90	1.40	2.00	3.70	1.07	1.50	2.13	4.00
1.7	0.90	1.50	2.13	4.10	1.07	1.70	2.45	4.40
1.8 ④	1.07	1.70	2.30	4.60	1.07	1.80	2.60	4.90

NOTES: (1) The distances specified in this table may be applied *only* where the transient overvoltage factor is known and is supplied by the employer.

(2) On systems where the voltage is controlled to operate at a maximum value in between the maximums listed in the table, clearance distances may be determined by interpolation, rounding to the next higher 0.1 m.

(3) There is no provision in OSHA 1926 for clearance reduction through application of transient overvoltage factors.

④ These distances are based on the highest transient overvoltage an employee will be exposed to on any system with live-line tools as the insulating material. If the transient overvoltage factor is not known, the values for this factor shall be used.

442. Switching Control Procedures

A. Designated Person

A designated person shall:

1. Keep informed of operating conditions affecting the safe and reliable operation of the system.
2. Maintain a suitable record showing operating changes in such conditions.
3. Issue or deny authorization for switching, as required, for safe and reliable operation.

B. Specific Work

Authorization from the designated person shall be secured before work is begun on or in the vicinity of station equipment, transmission, or interconnected feeder circuits and where circuits are to be de-energized at stations. The designated person shall be notified when such work ceases.

EXCEPTION: In an emergency, to protect life or property, or when communication with the designated person is difficult, because of storms or other causes, any qualified employee may make repairs on or in the vicinity of the equipment or lines covered by this rule without special authorization if the qualified employee can clear the trouble promptly with available help in compliance with the remaining rules. The designated person shall thereafter be notified as soon as possible of the action taken.

C. Operations at Stations

Qualified employees shall obtain authorization from the designated person before switching sections of circuits.

In the absence of specific operating schedules, employees shall secure authorization from the designated person before opening and closing supply circuits or portions thereof or starting and stopping equipment affecting system operation at stations.

EXCEPTION 1: Sections of distribution circuits are excepted if the designated person is notified as soon as possible after the action is taken.

EXCEPTION 2: In an emergency, to protect life or property, any qualified employee may open circuits and stop moving equipment without special authorization if, in the judgment of the qualified employee this action will promote safety, but the designated person shall be notified as soon as possible of such action, with reasons therefor.

D. Re-Energizing After Work

Instructions to re-energize equipment or lines which have been de-energized by permission of the designated person shall not be issued by the designated person until all employees who requested the line to be de-energized have reported clear. Employees who have requested equipment or lines de-energized for other employees or crews shall not request that equipment or lines to be re-energized until all of the other employees or crews have reported clear. The same procedure shall be followed when more than one location is involved.

E. Tagging Electric Supply Circuits

Equipment or circuits that are to be treated as de-energized shall have suitable tags attached to all points where such equipment or circuits can be energized. Controls that are to be de-activated during the course of work on energized or de-energized equipment or circuits shall also be tagged. The tags shall be placed to identify plainly the equipment or circuits on which work is being performed.

Tagging of Supervisory Control and Data Aquisition Systems (SCADA) in itself shall not be considered sufficient.

F. Restoration of Service After Automatic Trip

1. When controls upon which tags have been placed open automatically, they shall be left open until re-closing has been authorized.
2. When circuits open automatically, local operating rules shall determine in what manner and how many times they may be closed with safety.

G. Repeating Oral Messages

Each employee receiving an oral message concerning the switching of lines and equipment shall immediately repeat it back to the sender and obtain the identity of the sender. Each employee sending such an oral message shall require it to be repeated back by the receiver and secure the latter's identity.

443. Work on Energized Lines and Equipment

A. General Requirements

1. When working on energized lines and equipment, one of the following safeguards shall be applied:
   a. Insulate employee from energized parts
   b. Isolate or insulate the employee from ground and grounded structures, and potentials other than the one being worked on.

2. Employees shall not place dependence for their safety on the covering (non-rated insulation) of wires. All precautions (see Section 44) for working on energized parts shall be observed.
3. All employees working on or in the vicinity of lines or equipment exposed to voltages higher than those guarded against by the safety protective equipment provided shall assure themselves that the equipment or lines on which they are working are free from dangerous leakage or induction, or have been effectively grounded.
4. Cutting into Insulating Coverings of Energized Conductors
    a. A supply cable to be worked on as de-energized which cannot be positively identified or determined to be de-energized, shall be pierced or severed at the work location with a tool designed for the purpose.
    b. Before cutting into an energized supply cable the operating voltage shall be determined and appropriate precautions taken for handling conductors at that voltage.
    c. When the insulating covering on energized wires or cables must be cut into, the employee shall use a tool designed for the purpose. While doing such work, suitable eye protection and insulating gloves with protectors shall be worn. Employees shall exercise extreme care to prevent short circuiting conductors when cutting into the insulation.
5. Metal measuring tapes, and tapes or ropes containing metal threads or strands, shall not be used closer to exposed energized parts than the distance specified in Rule 441A. Care should be taken when extending metallic ropes or tapes parallel to and in the proximity of high voltage lines because of the effect of induced voltages.
6. Equipment or material of a noninsulating substance which is not bonded to an effective ground and extends into an energized area, and could approach energized equipment closer than the distance specified in Rule 441A, shall be treated as though it is energized at the same voltage as the line or equipment to which it is exposed.

B. Requirement for Assisting Employee

In inclement weather or at night, no employee shall work alone outdoors on or dangerously in the vicinity of energized conductors or parts of more than 750 V between conductors.

EXCEPTION: This shall not preclude a qualified employee, working alone, from cutting trouble in the clear, switching, replacing fuses, or similar work if such work can be performed safely.

C. Opening and Closing Switches

Manual switches and disconnectors shall always be closed by a continuous motion. Care should be exercised in opening switches to avoid serious arcing.

D. Working Position

Employees should avoid working on equipment or lines in any position from which a shock or slip will tend to bring the body toward exposed parts at a potential different than the employee's body. Work should, therefore, generally be done from below, rather than from above.

E. Protecting Employees by Switches and Disconnectors

When equipment or lines are to be disconnected from any source of electric energy for the protection of employees, the switches, circuit breakers, or other devices designated and designed for operation under the load involved at sectionalizing points shall be opened or disconnected first. When re-energizing, the procedure shall be reversed.

F. Making Connections

In connecting de-energized equipment or lines to an energized circuit by means of a conducting wire or device, employees should first attach the wire to the de-energized part. When disconnecting, the source end should be removed first. Loose conductors should be kept away from exposed energized parts.

G. Switchgear

Switchgear shall be de-energized prior to performing work involving removal of protective barriers unless other suitable means are provided for employee protection. The personnel safety features in switchgear shall be replaced after work is completed.

H. Current Transformer Secondaries

The secondary of a current transformer shall not be opened while energized. If the entire circuit cannot be properly de-energized before working on an instrument, a relay, or other section of a current transformer secondary circuit, the employee shall bridge the circuit with jumpers so that the current transformer secondary will not be opened.

I. Capacitors

Before employees work on capacitors, the capacitors shall be disconnected from the energizing source, short circuited, and grounded. Any line to which capacitors are connected shall be short circuited and grounded before it is considered de-energized. Since capacitor units may be connected in series-parallel, each unit shall be shorted between all insulated terminals and the capacitor tank before handling. Where the tanks of capacitors are on ungrounded racks, the racks shall also be grounded. The internal resistor shall not be depended upon to discharge capacitors.

J. Gas-Insulated Equipment

Employees working on gas-insulated cable systems or circuit breakers shall be instructed concerning the special precautions required for possible presence of arcing by-products of sulfur-hexafluoride (SF_6).

NOTE: By-products resulting from arcing in sulfur-hexafluoride (SF_6) gas-insulated systems are generally toxic and irritant. Gaseous by-products can be removed for maintenance on the compartments by purging with air or dry nitrogen. The solid residue that must be removed is mostly metallic fluoride. This fine powder absorbs moisture and produces fluorides of sulfur and hydrofluoric acid, which are toxic and corrosive.

K. Attendant on Surface

While electric supply personnel are in a manhole, an employee shall be available on the surface in the immediate vicinity to render assistance from the surface. This shall not preclude the employee on the surface from entering the manhole to provide short-term assistance.

EXCEPTION: This shall not preclude a qualified employee, working alone, from entering a manhole where energized cables or equipment are in service, for the purpose of inspection, housekeeping, taking readings, or similar work if such work can be performed safely.

L. Unintentional Grounds on Delta Circuits

Unintentional grounds on delta circuits shall be removed as soon as practical.

444. De-Energizing Equipment or Lines to Protect Employees

A. Application of Rule

1. When employees must depend on others to operate switches or otherwise de-energize circuits on which they are to work, or must secure special authorization before they operate such switches themselves, the precautionary measures that follow shall be taken in the order given before work is begun.
2. If the employee under whose direction a section of a circuit is disconnected is in sole charge of the section and of the means of disconnection, those portions of the measures that follow which pertain to dealing with the designated person may be omitted.
3. Records shall be kept on all contractual utility interactive systems on any electric supply lines. When these lines are de-energized according to Rule 444C, the utility interactive system shall be visibly disconnected from the lines.

B. Employee's Request

The employee in charge of the work shall apply to the designated person to have the particular section of equipment or lines de-energized, identifying it by position, letter, color, number, or other means.

C. Operating Switches, Disconnectors, and Tagging

The designated person shall direct the operation of all switches and disconnectors through which electric energy may be supplied to the particular section of equipment and lines to be de-energized, and shall direct that such switches and disconnectors be rendered inoperable and tagged. If switches that are controlled automatically, or remotely, or both can be rendered inoperable, they shall be tagged at the switch location. If it is impractical to render such switches and disconnectors inoperable then these remotely controlled switches shall also be tagged at all points of control. A record shall be made when placing the tag, giving the time of disconnection, the name of the person making the disconnection, the name of the employee who requested the disconnection, and the name, or title, or both, of the designated person.

D. Employee's Protective Grounds

When all the switches and disconnectors designated have been operated, rendered inoperable where practical, and tagged in accordance with Rule 444C, and the employee has been given permission to work by the designated person, the employee in charge should immediately proceed to make the employee's own protective grounds or verify that adequate grounds have been applied (see Rule 445) on the disconnected lines or equipment. During the testing for potential and/or application of grounds, distances not less than those shown in Tables 441-1 to 441-3, as applicable, shall be maintained.

Grounds shall be placed at each side of the work location and as close as practical to the work location, or a single point ground shall be placed at the work location. If work is to be performed at more than one location on a line section, the line section shall be grounded and short circuited at one location in the line section and the conductor to be worked on shall be grounded at each work location.

The distance in Tables 441-1, 441-2, or 441-3, as applicable, shall be maintained from ungrounded conductors at the work location. Where the making of a ground is impractical, or the conditions resulting therefrom are more hazardous than working on the lines or equipment without grounding, the ground may be omitted by special permission of the designated person.

E. Proceeding with Work

1. After the equipment or lines have been de-energized and grounded, the employee in charge, and those under the direction of the employee in charge, may proceed with work on the de-energized parts.

   Equipment may be re-energized for testing purposes only under the supervision of the employee in charge and subject to authorization by the designated person.

2. Each additional employee in charge desiring the same equipment or lines to be de-energized for the protection of that person, or the persons under direction, shall follow these procedures to secure similar protection.

F. Reporting Clear—Transferring Responsibility

1. The employee in charge, upon completion of the work

and after assuring that all persons assigned to this employee in charge are in the clear, shall remove protective grounds and shall report to the designated person that all tags protecting that person may be removed.

2. The employee in charge who received the permission to work may, if specifically permitted by the designated person, transfer the permission to work and the responsibility for persons by personally informing the affected persons of the transfer.

G. Removal of Tags

1. The designated person shall then direct the removal of tags and the removal shall be reported back to the designated person by the persons removing them. Upon the removal of any tag, there shall be added to the record containing the name of the designated person, or title, or both, and the person who requested the tag, the name of the person requesting removal, the time of removal, and the name of the person removing the tag.
2. The name of the person requesting removal shall be the same as the name of the person requesting placement, unless responsibility has been transferred according to Rule 444F.

H. Sequence of Re-Energizing

Only after all protective grounds have been removed from the circuit or equipment and after protective tags have been removed in accordance with Rule 444G at a specific location, may the designated person direct the operation of switches and disconnectors at that location.

445. Protective Grounds

A. Installing Grounds

When placing protective grounds on a previously energized circuit, the following sequence and precautionary measures shall be observed.

1. Size of Grounds

The grounding device shall be of such size as to carry the induced current and anticipated fault current that could flow at the point of grounding for the time necessary to clear the line.

2. Test for Voltage

   The previously energized parts which are to be grounded shall next be tested for voltage except where previously installed grounds are clearly in evidence. The employee shall keep every part of the body at the required distance by using insulating handles of proper length or other suitable devices.

3. Initial Connections

   The employee making a protective ground on equipment or lines shall first connect one end of the grounding device to an effective ground connection.

4. Completing Grounds

   If the test shows no voltage, or the local operating rules so direct, the grounding device shall next be brought into contact with the previously energized part using insulating handles or other suitable devices and securely clamped or otherwise secured thereto. Where bundled conductor lines are being grounded, grounding of each subconductor should be made. Only then may the employee come within the distances from the previously energized parts specified in Rule 441A or proceed to work upon the parts as upon a grounded part.

   NOTE: In stations, switches may be employed to connect the equipment or lines being grounded to the actual ground connection.

B. Removing Grounds

The employee shall first remove the grounding devices from the de-energized parts using insulating handles or other suitable devices.

Extreme caution shall be exercised that the proper sequence of installing or removing grounds is followed. The connection to the effective ground shall not be removed first. If done, electric shock and injury may result.

446. Live-Line Work

All employees using live-line work practices shall observe the following rules in addition to applicable rules contained elsewhere in Sections 42 and 44.

The distances specified in Tables 441-1, 441-2, or 441-3 shall be maintained from all grounded objects and from other conductors, lines, and equipment having a potential different from that to which conductive equipment and devices are bonded in order to maintain the equipotentially energized work environment in an isolated state.

A. Training

Employees shall be trained in live-line work practices which include rubber glove, hot stick, or bare-hand method, before being permitted to use these techniques on energized lines.

B. Equipment

1. Insulated aerial devices used in live-line work shall be tested before the work is started to assure the integrity of the insulation.
2. Insulated aerial devices and other equipment used in this work shall be maintained in a clean condition.
3. Tools and equipment shall not be used in a manner that will reduce the overall insulating strength of the insulated aerial device.

C. Bonding and Shielding for Bare-Hand Method

1. A conductive bucket liner or other suitable conducting device shall be provided for bonding the insulated aerial device to the energized line or equipment.
2. The employee shall be bonded to the insulated aerial device by use of conducting shoes, leg clips, or other suitable means.
3. Adequate electrostatic shielding shall be provided and used where necessary.
4. Before the employee contacts the energized part to be worked on, the aerial device shall be bonded to the energized conductor by means of a positive connection.

Appendix A

(This Appendix is not part of ANSI C2-1990, National Electrical Safety Code, 1990 edition, but is included for information only.)

Uniform System of Clearances Adopted in the 1990 Edition

Rules 232, 233, and 234

Introduction

The original format or system for stating NESC requirements was developed before 1920 and recognized the practical constraints of that time: Clearances were specified for a set of Basic Conditions. Some *Basic Clearances* included conductor movement; adders were used for non-basic conditions. Although easy to use, it was unduly conservative in many cases, and did not adequately recognize new materials and construction in others. Various additional clearance requirements were added over the years.

An intensive study by the NESC Clearances Subcommittee identified:

- modern utility practices and capabilities that remove the previous clearance measurement constraints,
- apparent inconsistencies in certain clearance treatments, and
- the need to develop a uniform clearance system independent of materials used for conductors and cables, stringing tensions, operating temperatures, and similar constraints.

The new uniform clearance system contained in the 1990 edition reflects the dimensions of expected activities in each area (reference component), as well as the relative potential problem caused by each type of facility (mechanical and electrical component).

Conductor clearance in the 1990 edition is stated in terms of the "closest approach," ie, the clear distance that must be maintained under specified conditions.

- Vertical clearances are required during maximum sag conditions; they provide for expected activity beneath a line.

- Horizontal clearances are required when the conductor is at rest; they provide for expected activity alongside a line. In addition, displacement of conductors by wind is considered under certain conditions.

Under the new system, users consider the actual characteristics of the materials and construction, rather than the reference characteristics built into the early code requirements.

While some clearance values in the new system may appear to be larger and some smaller, the net effective clearances for conductors and cables are, for most of the clearance values, essentially unchanged. Some few values required minor adjustments of the effective clearances to make them uniform with the other values, thus illustrating one of the needs for these changes.

The 1990 edition provides the following user benefits:

- simple code language in performance-standard format (as opposed to the prior design-manual style)
- readily understandable intent
- uniform clearance values
- integration of prior rules for long span construction and/or high conductor temperature operation
- reduced number of footnotes required to cover exceptions.

Clearance Rules and Tables Prior to 1990

In prior editions, clearances shown in the tables were *basic clearances,* applied under specified conditions of conductor temperature and sag, span length, and voltage range. For example, vertical clearances in Table 232-1 of the 1987 edition applied at a conductor temperature of 60 °F, no wind, final unloaded sag. Span lengths were limited by loading district. Voltages up to 50 kV were covered in the table. Conditions outside these basic conditions required additional clearances.

Actual clearances vary from the values required at 60 °F as conductor sag changes due to conductor movement under loading. Table 232-1 allowed 24 inches for ice loading, higher conductor temperatures (to 120 °F), and structure flexure. The actual allowance was 18 inches for ice loading or higher conductor temperature plus 6 inches for miscellaneous causes. Thus the true clearance requirement was 24 inches less than the value shown in the table (ie, equivalent to the clearance required for rigid parts).

Table 232-2, vertical clearance of *rigid* live parts, was consistent with the 24 inch allowance in Table 232-1. A rigid live part energized at 750 V to 22 kV over a road (item 1a, middle column) required 18 feet clearance. A conductor energized at the same voltage over a road required 20 feet clearance (Table 232-1, item 2). The additional 24 inches was the allowance for conductor movement, and the true clearance that may be experienced is 18 feet–the same as specified for rigid live parts.

Additional clearances above the basic values shown in Table 232-1 were required when the limiting conditions were exceeded. Rule 232B2c covered long span construction and Rule 232B2d covered high-temperature operation. Both rules recognized and allowed for additional conductor movement. Finally, Rule 232B1 required additional clearance for voltages exceeding 50 kV. Note that this was an electrical requirement apart from conductor movement.

Application of basic and additional clearances is illustrated in the drawings that follow.

Figure 1 shows the basic clearance applied at a 60 °F conductor temperature, at the limiting span length. When conductor movement is considered, the actual clearance at maximum sag is less than the basic clearance.

Figure 2 shows application of an additional clearance for conductor temperature over 120 °F, at the limiting span. Note that the additional clearance is added to the basic clearance to determine the required clearance. Again, the actual clearance at maximum sag is essentially the same as in Figure 1.

Figure 3 shows application of a long-span additional clearance, for a conductor operating at or within the 120 °F limit. As before, the additional clearance is added to the basic clearance, and the actual conductor clearance at maximum sag is essentially the same as in Figure 1.

To repeat, *both the basic clearances shown in the tables and the additional clearances required by the rules apply only when the conductor temperature is 60 °F. Actual clearances are expected and intended to be less due to conductor movement.* Only the voltage adder (for voltages above those shown in the tables) is a true clearance requirement.

Clearance Values Prior to 1990

Clearance values were based on experience. They were developed over time, at different times, for different reasons.

While those clearance values worked well, partly because several of them proved to be overly conservative, they did not follow a uniform system. For example, Table 234-1 showed vertical clearances from buildings. In line 1b(1) of the 1987 edition, vertical clearance of open supply conductors over roofs not accessible to pedestrians was the same for all voltages from 0 to 22 kV. However, in line 1b(3), the vertical clearance of open supply conductors over roofs accessible to vehicles but not subject to truck traffic varied with voltage: 12 feet for conductors energized at 0 to 300 V, 15 feet for 300 to 750 V, and 20 feet for 750 V to 22 kV.

Summary—Prior Editions

The clearance section was complex, with requirements stated in design manual format. Clearance was a mixture of basic clearance, clearance to cover conductor movement, and voltage clearance. Required clearance applied only at a 60 °F conductor temperature; actual clearance was not shown. Finally, clearance values were empirical; they were not systematized.

Clearances Subcommittee Activities

Subcommittee 4 discussed problems with the clearance section of the code during the 1987 revision cycle. Because there was insufficient time to develop a comprehensive proposal, Subcommittee 4 recommended formation of a special working group. The NESC Committee approved this recommendation and established Working Group 4.2 to:

- Review overhead line clearances, primarily Rules 232 and 234 and Section 28, and
- Investigate feasibility of a uniform method of determining clearances under all conditions of conductor movement.

The working group concluded that:

- A uniform system for determining clearances could be developed utilizing a building block approach.

- Vertical clearance values could be stated for maximum sag conditions to cover conductor movement.
- Horizontal clearances could be stated under at-rest conditions, with special requirements to cover displacement of energized conductors during wind conditions.
- The revisions proposed in the working group report were explicit, readily understood, and performance oriented.

Subcommittee 4 reviewed and approved the Working Group 4.2 report with minor modifications. Further enhancements were made in response to public comment.

The 1990 Changes

Rules 232, 233, and 234 were revised based on a coordinated, uniform system of clearances developed under a building block approach. Three components were considered to determine the total clearance required:

- *A reference component* to cover activity in the area to be cleared by the overhead supply and/or communication lines. For example, truck height for over-the-road transport is limited to 14 feet by state regulation. Thus the reference component for roads in Table 232-3 is 14 feet. Reference components included in the required clearances are shown in Table A-2.
- *A mechanical component* appropriate for the supply or communication line item. The mechanical component for open supply conductors is 2 feet (Table A-1).
- *An electrical component* appropriate for the voltage involved. The electrical component for open supply conductors, over 750 V to 22 kV, is 2.5 feet (Table A-1).

The required clearance is the sum of the three components: thus, 18.5 feet is required for open supply conductors, over 750 V to 22 kV, over roads (Table 232-1). For purposes of illustration, the mechanical and electrical components are combined in Table A-1, and items with the same total mechanical and electrical components are grouped into similar clearance categories. Six groups are thus created.

Application rules were revised to coordinate with clearances developed under the component or building block approach.

Vertical clearances now apply at the maximum conductor sag condition, such as outlined in Rule 232A, rather than at a 60 °F conductor temperature condition as used in the 1987 edition. This is illustrated in Figure 4: 18.5 feet is required for open supply conductors, over 750 V to 22 kv, over roads, for any sag condition or span length.

Not only is the intended vertical clearance shown, *rule simplification* is also achieved. Rules for long span construction and/or high temperature operation were removed because they are no longer necessary.

Horizontal clearances to buildings and other installations now apply with the conductor at rest (no wind displacement) as outlined in Rule 234A, rather than at a wind displacement condition. The horizontal clearance for open supply conductors from buildings (over 750 V to 22 kV) at rest is now 7.5 feet (Table 234-1). This clearance is essentially the same as the 8.0 feet required by early code editions.

Wind displacement need be considered only for energized open supply conductors and 230C2-230C3 cables energized at more than 750 V; see Rule 234C1. In the above example, a minimum clearance of 4.5 feet is required under wind displacement conditions. While less than the 5 or 6 feet required under prior codes, people will not be working on outside building walls during high wind conditions.

Because application rules have been revised, *it must be understood that clearance values cannot be directly compared between the 1987 and 1990 editions.* Vertical clearance values *appear* smaller because sag changes formerly included in clearance values are now addressed in the application rules. Horizontal clearance values *appear* larger because wind displacement is now applicable to energized conductors and certain supply cables only; clearances for all wires, conductors, and cables are shown in the tables under at-rest conditions.

The following changes were also made to consolidate requirements and simplify application:

- Voltages in the tables are limited to 0 to 750 V and over 750 V to 22 kV, normal secondary and primary distribution ranges respectively. Voltages in the 22 to 50 kV range are covered by a 0.4 inch per kV adder; see Rules 232C1a, 232C2a, and 234G1. Exceptions at 22 to 50 kV are noted where they apply.

- Rules for voltages above 22 kV and the alternate clearances for voltages above 98 kV are consolidated.
- Clearances for equipment cases are relocated from Rules 286E and 286F to Rules 232B3 and 234J.

Summary

These changes constitute a comprehensive revision of Section 23 which incorporate related provisions of Section 28 and provide significant user benefits, as detailed in the introduction at the beginning of this discussion. While some clearance values may appear to be larger and some smaller, the net effective clearances for energized conductors and cables are essentially unchanged.

Table A-1

Group	M&E (ft)	Category	R/NR	GI/O	M	E
I	1.0	Support Arms	1.0/—	0.0/—	1.0	0.0
		Effectively Grounded Equipment Cases	1.0/—	0.0/—	1.0	0.0
II	1.5	Insulated Communication Conductors and Cables	—/1.5	0.0/—	1.5	0.0
		Messengers	—/1.5	0.0/—	1.5	0.0
		Surge Protection Wires	—/1.5	0.0/—	1.5	0.0
		Grounded Guys	—/1.5	0.0/—	1.5	0.0
		230E1	—/1.5	0.0/—	1.5	0.0
		230C1	—/1.5	0.0/—	1.5	0.0
III	2.0	URLP, 0 - 750 V	1.0/—	—/0.5	1.5	0.5
		Non-Insulated Communication Conductors	—/1.5	—/0.5	2.0	0.0
		230C2, 0 - 750 V	—/1.5	0.0/—	1.5	0.5
		230C3, 0 - 750 V	—/1.5	0.0/—	1.5	0.5
		Ungrounded Cases of Equipment at 0 - 750 V	1.0/—	—/0.5	1.5	0.5
IV	2.5	230C2, Greater Than 750 V	—/1.5	0.0/—	1.5	1.0*
		230C3, Greater Than 750 V	—/1.5	0.0/—	1.5	1.0*
		Open Supply Conductors, 0 - 750 V	—/1.5	—/0.5	2.0	0.5
V	4.0	URLP, Greater Than 750 V — 22 kV	1.0/—	—/0.5	1.5	2.5
		Ungrounded Cases of Equipment at Greater Than 750 V — 22 kV	1.0/—	—/0.5	1.5	2.5
VI	4.5	Open Supply Conductors, Greater Than 750 V — 22 kV	—/1.5	—/0.5	2.0	2.5

LEGEND:

URLP — Unguarded Rigid Live Parts
R — Rigid = 1.0 ft
NR — Non-Rigid = 1.5 ft
GI — Grounded or Insulated = 0.0 ft
O — Bare, Ungrounded or Open Conductor or Part = 0.5 ft
M — Mechanical Component = R/NR plus GI/O
E — Electrical Component
- Grounded & Communications Conductor = 0.0 ft
- Supply Line 0-750 V = 0.5 ft
- Supply Line Greater Than 750 V — 22 kV = 2.5 ft

M&E — Sum of M and E values

NOTES: (1) Ungrounded guys have clearances based on the highest voltage to which they are exposed.
(2) An asterisk (*) beside a value indicates an exception to the legend.

Table A-2a
Reference Components of Rule 232

	Table 232-1		Table 232-2	
	Item	Ref (ft)	Item	Ref (ft)
Track rails	1	22.0	—	—
Roads, streets, alleys, etc.	2	14.0	1a	14.0
Residential driveways, etc.	3	14.0	1b	14.0
Other land traversed by vehicles	4	14.0	1c	14.0
Spaces and ways — pedestrians	5	8.0/10.0	1d	10.0
Water areas — no sailboating	6	12.5	—	—
Water areas — sailboating	7		—	
(a) Less than 20 acres		16.0		—
(b) Over 20 to 200 acres		24.0		—
(c) Over 200 to 2000 acres		30.0		—
(d) Over 2000 acres		36.0		—
Areas posted for rigging or launching sailboats	8	See 7	—	
Within or along:				
Roads, streets, or alleys	9	14.0	2a	14.0
Rural districts, vehicles unlikely	10	12.0	2b	12.0

Table A-2b
Reference Components of Rule 234

Table	Item	Ref (ft)
234-1	1. Buildings	
	a. Horizontal	
	(1) Walls, projections and guarded windows	3.0
	(2) Unguarded windows	3.0
	(3) Balconies and areas accessible to pedestrians	3.0
	b. Vertical	
	(1) Roofs/projections not accessible to pedestrians	8.0
	(2) Balconies and roofs accessible to pedestrians	9.0
	(3) Roofs — vehicles not over 8 feet	9.0
	(4) Roofs — vehicles over 8 feet	14.0
	2. Signs, chimneys, billboards, antennas, tanks, etc.	
	a. Horizontal	3.0
	b. Vertical over or under	3.5
234-2	1. Over bridges	
	a. Attached	1.0
	b. Not attached	8.0
	2. Beside, under or within bridge structure	
	a. Accessible	
	(1) Attached	1.0
	(2) Not attached	3.0
	b. Inaccessible	
	(1) Attached	1.0
	(2) Not attached	2.0
234-3	A. From water level, edge of pool, etc.	20.5
	B. From diving platform or tower	12.5

Appendix A

Prior Method

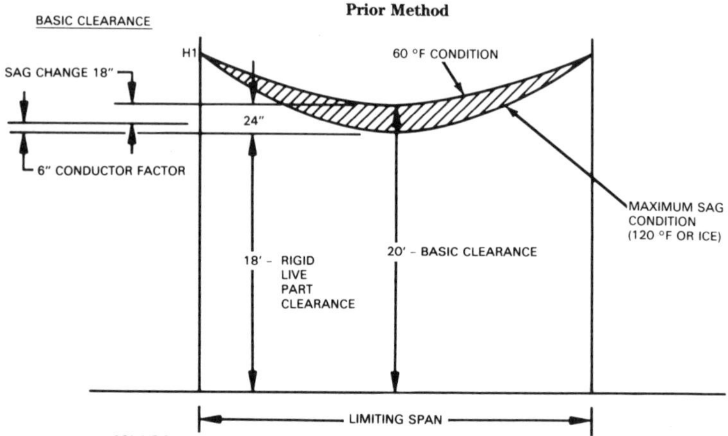

CONDITION — CLEARANCE OF OPEN SUPPLY CONDUCTORS, 750 V–22 kV, ABOVE ROADS
LIMITING SPAN — 175' MAXIMUM IN HEAVY LOADING DISTRICT
/////// — CONDUCTOR MOVEMENT

BASIC CLEARANCE (20') = ACTUAL = ACTUAL CLEARANCE (18') + CONDUCTOR MOVEMENT (2')

Fig 1
Basic Clearance

Prior Method

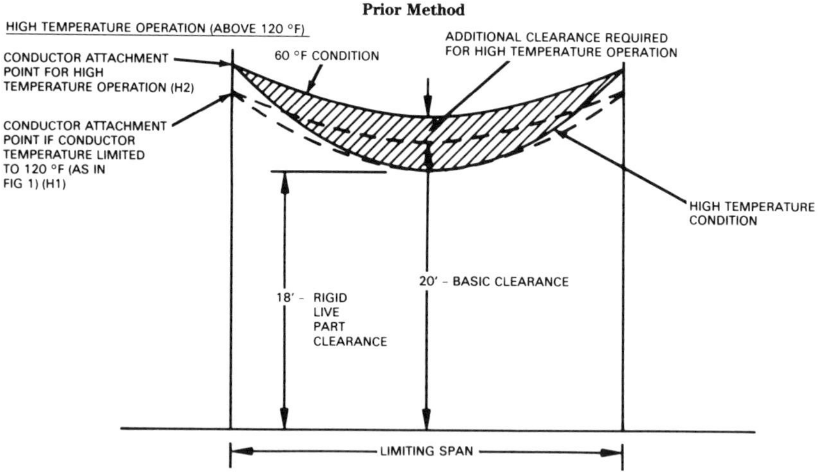

CONDITION — CLEARANCE OF OPEN SUPPLY CONDUCTORS, 750 V–22 kV, ABOVE ROADS
/////// — CONDUCTOR MOVEMENT
----- POSITION OF CONDUCTOR FOR BASIC CLEARANCE CONDITION (FIG 1)

REQUIRED CLEARANCE (@ 60 °F) = BASIC CLEARANCE (20') + ADDITIONAL CLEARANCE REQUIRED FOR HIGH TEMPERATURE OPERATION (ABOVE 120 °F)

Fig 2
Additional Clearance Required for High Temperature Operation

Appendix A

Prior Method

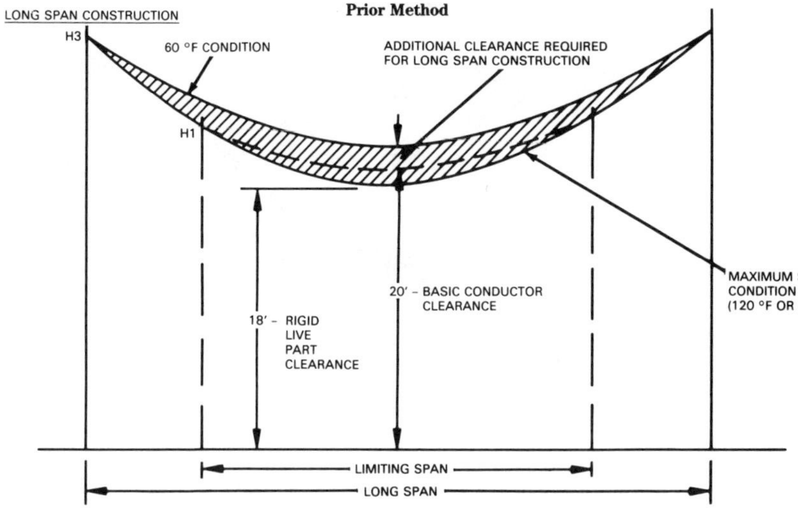

CONDITION — CLEARANCE OF OPEN SUPPLY CONDUCTORS, 750 V-22 kV, ABOVE ROADS

/////// — CONDUCTOR MOVEMENT

----- POSITION OF CONDUCTOR FOR BASIC CLEARANCE CONDITION (FIG 1)

REQUIRED CLEARANCE (@ 60 °F) = BASIC CLEARANCE (20′) + ADDITIONAL CLEARANCE REQUIRED FOR LONG SPAN CONSTRUCTION

(ASSUMING MAXIMUM CONDUCTOR TEMPERATURE LIMITED TO 120 °F)

Fig 3
Additional Clearance Required for Long Span Construction

1990 Method

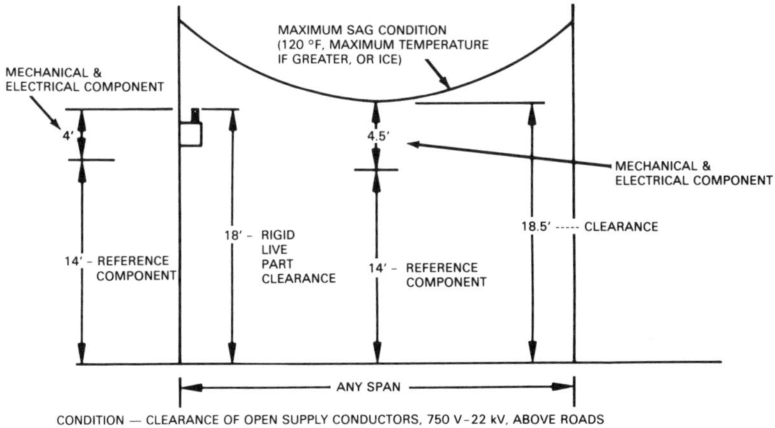

CONDITION — CLEARANCE OF OPEN SUPPLY CONDUCTORS, 750 V-22 kV, ABOVE ROADS

CLEARANCE (18.5′) = REFERENCE COMPONENT (14′) + MECHANICAL AND ELECTRICAL COMPONENT (4.5′)

Fig 4
Clearance at Maximum Sag

Index

Italic type is used for C2 *rule, section, part, figure,* or *table* identification. The corresponding page numbers follow in upright type.

A

AC live-line work minimum clearance, *Table 441-1;* 373
With surge factor, *Table 441-2;* 374, 376

Accessibility
Safety rules, *213;* 139
Switches, *216;* 142

Adjustable-speed motors, *130E;* 126

Administrative authority, *Sec. 2;* 50

Aerial cable systems, *278;* 324
Electrical requirements, *278A;* 324
Mechanical requirements, *278B;* 324

Alternating current circuits
Clearance, *230G;* 155
Ground connections, *92B;* 67

Ampacity, grounding conductor, *93C;* 72

Anchor rod (guy), *264G;* 321

Angles in line, strength requirements, *261A1b, 261A2a(4);* 286, 288

Application of rules, *013;* 46

Armless construction
Brackets, grades of construction for, *243C;* 275
Strength requirements, *261G;* 297

Assumed transverse loading, *252B;* 281

Assumed vertical loading, *252A;* 281

Automatic, *Sec. 2;* 50

B

Backfill, *Sec. 2;* 50
Underground conduits, *321;* 333

Ballast section, *Sec. 2;* 50

Bare grounded conductor, *354C2;* 350

Boilers (fuel oil fired), *127F;* 113

Bonding, *Sec. 2;* 50
Cable in underground structures, *342;* 343
Direct buried cable, *350E;* 345
Equipment frames and enclosures, *93C7;* 74

Bonding *(contd)*
Grounding electrodes (communication circuits), *99C;* 87

Boring, *353C;* 348

Bracing, *264;* 320
Point of attachment, *264C;* 320
Strength requirements, *261D4b, 264B;* 295, 320
Where used, *264A;* 320

Brackets
Clearances for, *238C;* 254
Strength requirements, *261D4c, 262E;* 295, 310
Vertical clearance from communication lines, *Table 238-2;* 256

Bridges
Supporting structures attached to, *217A1d;* 143
Underground conduits, *320A4, 322B5;* 331, 334

Buckarm construction, climbing space on, *236F;* 246

Buckarms
Location relative to working spaces, *237D;* 252
Obstruction of working space by, *Fig 237-1;* 254

Buildings (and other structures), direct buried cable, *351C2;* 346

Building walls, underground conduits through, *322B4;* 334

Bulk plants, classified areas, *Table(s) 127-2, 127-3;* 119

Buried systems. *See* Direct buried cable, Underground lines

Bus
Metal enclosed, *181;* 136
Isolated phase, *181B;* 136

C

Cable, *Sec. 2;* 50. *See also* Conductors, Direct buried cable, Wire
Accessories and joints, *333;* 341
Conduits or coverings, *239G3;* 254
Fiber optic, *Sec. 2,* 54
Clearance, *230F;* 154
Fireproofing, *Sec. 2;* 54
Insulation
Metal-sheathed, *163B;* 130
Shielded, *163B;* 130
Jacket, *Sec. 2;* 50
Jacketed multiple-conductor, *239G2;* 254
Meeting Rule 230C1, *239G1;* 260
Metal-sheathed communication cables, *239H1;* 262
Sheath, *Sec. 2;* 51
Supply, clearance, *230C;* 153
Terminal, *Sec. 2;* 51

Cable in underground structures, *Sec. 34;* 341
Communication cables containing special supply circuits, *344;* 344
Fireproofing, 343; 344
General, *340;* 341
Grounding and bonding, *342;* 343
Installation, *341;* 341
Cable in manholes and vaults, *341B;* 342
General, *341A;* 341

Capacity (rating)—switches, *173A;* 132

Circuit breakers, *Sec. 2, Sec. 17;* 51, 131
Application, *171;* 131
Arrangement, *170;* 131
Oil in, *172;* 131

Circuits, *Sec. 2;* 51
Alternating current clearance requirements, *230G;* 155
Common neutral (grounding), 215B1; 140
Constant-current clearance, *230H;* 155

Circuits *(contd)*
Current-transformer secondary circuits protection, *150;* 128
Direct current clearance requirements, *230G;* 155
Earth as part of, *215B5; 314C4;* 140, 329
Grounding, *215;* 140
Secondary, grounding, *151;* 128

Circuits of one class used exclusively in operation of circuits of another class, *224;* 149
Used in operation of communication circuits, *224B;* 150
Used in operation of supply circuits, *224A;* 149

Classified locations, *127;* 107
Bulk plants (equipment), *Table(s) 127-2, 127-3;* 119
Flammable liquid storage areas, *Table 127-1;* 116
LPG storage, Table *127-4;* 122
Natural gas (methane), *Table 127-5;* 124

Clearance, *Sec. 2, Sec. 23;* 51, 153
Additional clearances for voltages exceeding 22 kV for wires, conductors, cables, *234G;* 200
Alternate clearances for voltages exceeding 98 kV ac to ground or 139 kV dc to ground, *234H;* 201
Alternate clearance to surface of support arms and structures, *235E3b;* 228
In any direction from line conductors to supports; *Table 235-7;* 244
In any direction from line conductors to supports and to vertical or lateral conductors, span, or guy wires attached to same support, *Table 235-6;* 240, 242
Clearance diagram
For building, *Fig 234-1(a);* 204
For energized conductor, *Fig 235-1;* 231
For other structures, *Fig 234-1(b);* 205
Clearance envelope, *Fig 233-3;* 183
And conductor movement envelope to determine applicable clearance, *Fig 233-1;* 181
For grain bins, *Fig 233-4;* 207
Conductor movement envelope, *Fig 233-2;* 182, 183
Constant-current circuits, 155
Direct current circuits, *230G;* 155
Electrical clearances in Rules 235B3a(l), *Table 235-4;* 235
Electrical component of, *233C3b, 234H3;* 180, 201
Electrical component to buildings, bridges, and other installations in Rule 234G3a, *Table 234-4;* 218
Of equipment mounted on supporting structures, *234J;* 203
General, *230;* 153
Horizontal, *233B, 234A1, 234C1, 234D1;* 178, 193, 194, 197
Alternate clearance for voltage exceeding 98 kV ac to ground or 139 kV dc current to ground, *233B2;* 178
Clearance requirements, *233B1;* 178
Between conductors bounding climbing space, *Table 236-1;* 250
Between line conductors, *235B;* 220
Between line conductors AWG no. 2 or larger at supports, based on sags, *Table 235-3;* 233, 234
Between line conductors smaller than AWG no. 2 at supports, based on sags, *Table 235-2;* 233

Clearance *(contd)*

Horizontal *(contd)*

Transition between vertical and, *234A3;* 193

With wind displacement, *234A2, 234C1b, 234D1b;* 193, 194, 197

Between wires, conductors, or cables support, *Table 235-1;* 232

Lateral conductors, *Table 239-1;* 264, 265

Lateral facilities from other facilities and surfaces on same support, *239;* 257

Maintenance, *230;* 155

Measurement, *230B;* 153

Between open vertical conductors and pole center, *Table 239-2;* 264, 265

Rail car clearances, *Fig 234-3;* 206

Between supply circuits of different voltage classifications on same support arm, *235F;* 229

Of supply conductors and cables from bridges, *Table 234-2;* 212

Between supply wires, conductors, and cables in Rule 233C3b(l), *Table 233-2;* 191

Of supply wires, conductors, cables, and rigid live parts from bridges, *234D;* 197

Vertical, *233C, 234A1, 234C1, 234D1;* 179, 193, 194, 197

Above ground wires, conductors, cables, roadways, rails, or water surfaces, *232, Table 232-1;* 156, 161, 166

Application, *232A;* 157

On supporting structures, *232B;* 157

AC for voltage exceeding 98 kV ac to ground or 139 kV dc to ground, *233C3;* 179

Between certain communication and supply facilities located on same structure, *238;* 254

Clearance requirements, *233C1;* 179

Between communication conductors and supply equipment, *Table 238-1;* 255

Between conductors at supports, *Table 235-5;* 236, 238

Electrical component of clearance in Rule *232D3a;* 176

Between line conductors, *235C;* 224

Reference heights, *Table 232-3;* 175

Of rigid live parts above ground or roadway surfaces, *Table 232-2;* 171, 173

Of span wires and brackets from communication lines, *Table 238-2;* 256

Between supply and communication equipment, *Table 238-1;* 255

Between supply conductors and communication equipment, *Table 238-1*; 255

Transition between horizontal and, *234A3;* 193

Voltages exceeding 98 kV, *232D;* 158

Voltages exceeding 22 kV, *232C1, 232D;* 158, 179

Between wires, conductors, and cables on different supporting structures, *Table 233-1;* 184, 188

Of vertical conductors, *Table 239-1;* 264, 265

Of vertical facilities from other facilities and surfaces on same support, *239;* 257

Between wires, conductors, and cables carried on different support structures, *233;* 177

Of wires, conductors, and cables from buildings, bridges, rail cars, swimming pools, and other installations, *234;* 193

Clearance *(contd)*
Vertical *(contd)*
Of wires, conductors, and cables passing by but not attached to buildings, *Table 234-1;* 208, 210
Of wires, conductors, and cables passing over or near swimming areas, *Table 234-3;* 216
Of wires, conductors, cables, and rigid live parts from buildings, signs, billboards, chimneys, radio and television antennas, tanks, and other installations except bridges, *234C;* 194
For wires, conductors, or cables carried on same supporting structure, *235;* 220
Of wires, conductors, or cables over or near swimming areas, *234E;* 198

Climbing space, *236;* 245
Between conductors, *236E;* 245
On Buckarm construction, *236F;* 245
Horizontal clearance between conductors bounding, *Table 236-1;* 250
Location and dimensions, *236A;* 245
Location of equipment relative to, *236D;* 245
Location of vertical or lateral conductors relative to, *239B;* 257
Past longitudinal runs not on support arms, *236G, Fig 236-1;* 246, 248
Past vertical conductors, *236H;* 247
Portions of supporting structures in, *236B;* 245
Near ridge-pin conductors, *236I;* 247
Support arm location relative to, *236C;* 245

Coal handling areas, *127A;* 107

Combustible liquids, *127B;* 110

Common
Grounding conductor, *93F;* 76
Neutral circuits, *215B1;* 140
Use, *Sec. 2;* 51

Communication
Lines, *Sec. 2;* 51, 55
Protective requirements, *315;* 329

Communication cables
Metal-sheathed, *239H1;* 262
Random separation in direct burial, *354B, 354C;* 350
Strength requirements, *261K;* 299

Communication circuit conductors, *242C;* 268

Communication conductors, *239H2;* 262
Grade N construction, *263G;* 318
Grades of communication for communication conductors alone, or in upper position of crossing or on joint poles, *Table 242-2;* 273

Communication employee rules
Approach to energizing conductors or parts, *431;* 371
Attendant on surface at joint-use manhole, *433;* 372
Climbing joint-use poles, *432;* 371
Sheath continuity, *434;* 372

Communication grounding conductors, *239H3;* 263

Communication lines, safety rules, *Part 2;* 138

Concentric neutral cable, *94B5;* 80

Concrete crossarms, *261D2;* 294

Concrete-encased (grounding) electrodes, *94B6;* 80

Conductor fastenings
Grade of construction, *243C;* 275
Strength requirements, *261F2;* 296

Conductor(s), *Sec. 2, Sec. 16, 262H;* 51, 129, 311. *See also* Cable, Wire
Bundled, *Sec. 2;* 51
Climbing space between, *236E;* 245
Communication (Grade N construction), *263G;* 318
Communication conductors, *239H2;* 262
Communication grounding conductors, *239H3;* 263
Conduits or coverings, *239G3;* 254
Constant-current circuit conductors, *242A;* 268
Covered, *Sec. 2;* 51
Clearance, *230D;* 154
Clearance requirements, *230D;* 154
Electrical protection, *160;* 129
Existing installations, *013B3;* 46
Fiber optic, *Sec. 2;* 51
Grades of construction, *242;* 268
Grounded, *Sec. 2;* 51
Grounding, *Sec. 2;* 51
Ampacity, *93C;* 72
Common, *93F;* 76
Guarding, *93D;* 75
Protection, *93D;* 75
Separation of circuits, *97;* 84
Surge arresters, *192;* 137
Underground, *93E;* 75
Guys, *92C;* 69
Insulated, *Sec. 2;* 52
Isolation, *162;* 130
Lateral, *Sec. 2;* 52
Line, *Sec. 2;* 52
Loading, *251;* 280
Mechanical protection, *161;* 130
Messenger wires, *92C;* 69
Of more than 300 volts to ground, *234C3a;* 196
Neutral, *230E;* 154
Not in conduit, *239C;* 257
Open, *Sec. 2;* 52
Paired conductors without messengers, *262H2;* 311
Random separation in direct burial, *354A, 354B, 354C;* 349, 350
Relatives for various classes of lines, 220; 144
Ridge-pin, climbing space near, *236I;* 247
Sag, *262H3;* 311
Separation, *97;* 84
Service-drop, *234C3c;* 196
Shielding, *Sec. 2;* 52
Size, *Table 261-7, 262H1;* 306, 311
Spacing, *235G;* 230
Splices, *262H4;* 311
Supply, *261H1;* 297
Supply grounding conductors, *239G5;* 261
Supply line, grade N, sizes for, *263D, Table 263-1;* 317, 319
Terminations, *163;* 130
Insulation, *163A;* 130
Metal-sheathed or shielded cable, *163B;* 130
Vertical, *236H;* 247
Vertical clearances, *232;* 156

Conduit, *Sec. 2;* 52

Conflict, avoidance of, *221;* 148

Constant-current circuit conductors, *242A;* 268

Constant-current circuits, clearance, *230H;* 155

Construction, grades of. *See* Grades of construction

Control
Motors, *131;* 126
Switchboards, *180E;* 135

Control circuits, *130F;* 126

Crossarm cross section, dimensions of, *Table 261-6;* 305

Crossarms
Dimensions of cross section, *Table 262-4:* 316
Double crossarms, *262D2;* 310
Grade of construction, *243B;* 274
Material and size, *262D1;* 310
Strength requirements, *261D;* 293
Grade N construction, *263C;* 317

Crossings
Direct underground cable, *352B;* 347
Grades of construction at, *241C;* 266
Multiple, *241C3;* 267

Current-carrying part, *Sec. 2;* 52

Cylindrical structures and components, *252B2a;* 282

D

DC minimum clearance for live parts, *Table 441-3;* 378, 379

Dead end
Longitudinal loading, *252C3;* 284
Strength requirements, *261A2a(3);* 287
Open supply conductors, *261H3;* 298
Pin construction, *261F1c;* 296

Dead front power switchboards, *180C;* 135

De-energized, *Sec. 2;* 52

De-energizing equipment or lines, *444;* 385
Application of Rule *444A;* 385
Employee's protective grounds, *444D;* 386
Employee's request, *444B;* 385
Operating switches, disconnectors, and tagging, *444C;* 385
Proceeding with work, *443E;* 386
Removal of tags, *443G;* 387
Reporting clear-transferring responsibility, *443F;* 386
Sequence of re-energizing, *443H;* 387

Definitions, *Sec. 2;* 50

Designated fiber stress, *261A2a(1);* 287

Designated person, *Sec.* 2; 52

Direct buried cable, *Sec. 35;* 345
Depth of burial, *353D;* 349
General, *350;* 345
Installation, *353;* 348
Location and routing, *351;* 345
General, *351A;* 345
Natural hazards, *351B;* 346
Other considerations, *351C;* 346
Protection, *354E;* 352
Random separation, *354;* 349
Separations from other underground structures, *352;* 347
Supply cable or conductor burial depth, *Table 353-1;* 349

Direct current circuits, *230G;* 155

Direct current systems
Grounded connections, *92B;* 67
Grounding point, *92A;* 67

Disconnecting
Devices, *173;* 132
Switch, *Sec. 2;* 53

Disconnection fuses, *174;* 132

Double crossarms, *262D2;* 310

Drainage, *323G;* 338

Drip loops of luminaire, *238D;* 254

Ducts, *Sec. 2;* 53
Underground conduits, *322;* 333

E

Effective date, *016;* 48

Effectively grounded, *Sec. 2;* 53

Electric railway construction, *225;* 151

Electric supply
Equipment, *Sec. 2;* 53
Lines, *Sec. 2;* 53, 56
Stations, *Sec. 2;* 53
Exits, *113;* 93
Fire extinguisher equipment, *114;* 94
Floor, floor openings, passageways, stairs, *112;* 93

Electric supply *(contd)*
Stations *(contd)*
Installation and maintenance, *Part 1;* 88
Purpose of rules, *100;* 88
Scope of rules, *101;* 88
Protective arrangements, *Sec. 11;* 88
General requirements, *110;* 88

Electrode
Communication circuits, *99;* 86
Grounding, *94;* 76
Method of connection, *95;* 81
System (grounding) isolation, *97E;* 86

Electrolysis, *264F;* 320

Emergency installations, *014;* 46-47

Employers, rules for, Sec. 41; 359. *See also* Communication employee rules
Arcing conditions, *420E;* 364
Cable reels, *420O;* 366
Clothing, *420I;* 364
Emergency procedures and first aid rules, *410B;* 360
Energized or unknown conditions, *420D;* 363
Fire extinguishers, *420L;* 365
Fuses, *420N;* 366
General requirements, *410;* 359
Ladders and supports, *420J;* 364
Liquid cell batteries, *420G;* 364
Machines or moving parts, *420M;* 365
Personal general precautions, *420;* 362
Qualifications of employees, *420B;* 362
Responsibility, *410C;* 360
Rules and emergency methods, *420A;* 362
Safeguarding oneself and others, *420C;* 363
Safety straps, *420K;* 365
Street and area lighting, *420P;* 366
Tools and protective equipment, *420H;* 364
Ungrounded metal parts, *420E;* 364

Enclosed, *Sec. 2;* 53

Enclosures
Drains, *181B3;* 136
Electrical clearances, *373;* 354
Live parts, *124C2;* 97

Energized, *Sec. 2;* 53

Energized lines and equipment, *443;* 381
Attendant on surface, *443K;* 384
Capacitors, *443I;* 384
Current transformer secondaries, *443H;* 383
Gas-insulated equipment, *443J;* 384
General requirements, *443A;* 381
Making connections, *443F;* 383
Opening and closing switches, *443C;* 383
Protecting employees by switches and disconnectors, *443E;* 383
Requirement for assisting employee, *443B;* 383
Switchgear, *443G;* 383
Unintentional grounds on delta circuits, 443L; 384
Working position, *443D;* 383

Energized parts, *126;* 107

Equipment, *Sec. 2, Sec. 38;* 53, 355
Cases
Clearance to, *234J;* 203
Vertical clearance, *232B3;* 157
Classified areas
Bulk plants, *Table 127-2, Table 127-3;* 119
Flammable liquid storage areas, *Table 127-1;* 116
Liquid petroleum gas, *Table 127-4;* 122
Natural gas (Methane), *Table 127-5;* 124
Design, *381;* 355
Employee rules for, *420H;* 364
Enclosure, *110A;* 88

Equipment *(contd)*
Energized
Guarding, *237E;* 253
Working clearances from, *237F;* 253
Existing installations, *013B3;* 46
Gaseous hydrogen systems, *127E;* 113
General, *380;* 355
Grounding, *215;* 140
Guarding, *237E;* 253
Identification, *128, 385;* 125, 357
Inspection and tests of, *214;* 139
When in service, *214A;* 139
When out of service, *214B;* 140
Installation, *383;* 357
Installation and maintenance, *Sec. 12,* 94
General requirements, *120;* 94
Guarding live parts, *124;* 96
Guarding shaft ends, pulleys, belts, and suddenly moving parts, *122;* 95
Inspections, *121;* 94
Live parts, *232B2;* 157
Location
Relative to climbing space, *236D;* 245
In underground structures, *382;* 356
Mobile hydrogen supply units, *129;* 125
Pad-mounted, *Sec. 2;* 56
Rooms and spaces, *110B;* 89
Rotating, *Sec. 13;* 125
Rules, *Sec. 10;* 88
Purpose, *100;* 88
Scope, *101;* 88
Support for stationary equipment, *110C;* 89
Unguarded rigid live parts of, *234J1;* 203
Vertical clearance between certain communication and supply facilities, *238A;* 254
For work on energized parts, *126;* 107

Examples, *015D;* 47

Excavation, underground conduits, *321;* 333

Existing installations, *013B;* 46

Explosion-proof apparatus, *Sec. 2;* 53

Exposed, *Sec. 2;* 53

Extensions, *013A;* 46

F

Fences, grounding, *92E, 93C6;* 70, 74

Fiber optic cable
Clearance, *230F;* 154
Communication, *Sec. 2;* 54
Supply, *Sec. 2;* 54

Fire-alarm circuit conductors, *242D;* 268

Fire hydrants, clearance requirements, *231A;* 155

Fireproofing, *Sec. 2;* 54
Cables in underground structures, *343;* 344

Fixed supports
Clearance, *235E1;* 227
Horizontal clearance, *235B1;* 220

Flammable liquids, *127B;* 110
Storage area, *127C;* 110
Classified areas, *Table 127-1;* 116

Flat surface structures and components, loading, *252B2b;* 282

Flexible supports, strength requirements, *261A4b;* 292

Floors, battery areas, *144;* 127

Foundations
Overload capacity factors, *Table 261-4;* 305
Strength requirements, *261B;* 292

Fuel lines, underground conduits, *320B5;* 333

Fuses, *Sec. 17;* 131
Application, *171;* 131
Arrangement, *170;* 131
Disconnection, *174;* 132

G

Gaseous hydrogen systems (Supply equipment), *127G;* 113
Gasoline dispensing stations, *127E;* 112
Classified areas (Equipment), *Table 127-3;* 119
General operating routines, *421;* 366
Area protection, *421B;* 366
Duties of person in charge, *421A;* 366
Escort, *421C;* 367
General rules, *012;* 45
Grade D construction, *262;* 307
Grade N construction, *263;* 317
Grades B and C construction, *261;* 286
Grades of construction, *Sec. 24;* 266
Armless construction brackets, *243;* 275
Assumed longitudinal loading, *252C1;* 283
Communication circuit conductors used exclusively in operation of supply lines, *242C;* 268
For communication conductors alone, or in upper position of crossing or on joint poles, *Table 242-2;* 273
For communication conductors crossing over railroad tracks and supply lines, *Table 241-1;* 267
Conductor fastenings, *243C;* 275
For conductors, *242;* 268
Conflicts, *241D;* 267
Constant-current circuit conductors, *242A;* 268
Crossarms, *243B;* 274
At crossings, *241C;* 266
Fire-alarm circuit conductors, *242D;* 268
General, *240;* 266
Insulators, *243C;* 275
For line supports, *243;* 274
Neutral conductors of supply circuits, *242E;* 268
Order of grades, *241B;* 266
Pins, *243C;* 275
Railway feeder circuit conductors, *242B;* 268
Structures, *243A;* 274
Supply cables, *241A;* 266
For supply conductors alone, at crossing, or on same structures with other conductors, *Table 242-1;* 270
Support arms, *243B;* 274
Surge protection wires, *242F;* 269
Trolley-contact circuit conductors, *242B;* 268

Grain bins
Clearance envelope, *Fig 234-4;* 207
Clearance of wires, conductors, cables, and rigid live parts from, *234F;* 199

Ground
Connection, *95A;* 81
Contact surfaces, *95C;* 83
Piping system, *95B;* 82
Fault interrupter (GFI), *111E;* 90
Resistance, *96A;* 83
Multi-grounded systems, *96A3;* 84
Single-grounded systems, *96A2;* 84
Supply stations, *96A1;* 83

Grounded, *Sec. 2;* 54
Effectively, *Sec. 2;* 54
System, *Sec. 2;* 54

Grounding, *Sec. 9;* 65
Alternating-current systems, *92B;* 67
Cable in underground structures, *342;* 343

Grounding *(contd)*
Circuits, *215;* 140
Communication grounding conductors, *239H3;* 263
Conductor, *93A;* 67
Ampacity, *93C;* 72
Limit, *93C8;* 74
Bonding of equipment frames and enclosures, *93C7;* 74
Common, *93F;* 76
Composition, *93A;* 71
Current in, *92D;* 70
Equipment, *93C5;* 73
Guarding, *93D;* 75
Instrument transformer, *93C3;* 73
Point of connection, *92;* 67
Primary surge arrester, *93C4;* 73
Protection, *93D;* 75
Separation of circuits, *97;* 84
Strength, *93C, 93C9;* 72, 74
Surge arresters, *192;* 137
Underground, *93E;* 75
Direct-current systems, *92A;* 67
Electrodes, *94;* 76
Existing, *94A;* 77
Local systems, *94A2;* 77
Metallic water piping system, *94A1;* 77
Steel reinforcing bars, *94A3;* 77
Made, *94B;* 77
Buried wire, strips, plates, *94B3;* 78
Driven rods, *94B2;* 78
General, *94B1;* 77
Pole butt plates and wire wraps, *94B4;* 79
Method of connection, *95;* 81
Equipment, *215, 384;* 140, 357
Fences, *92E;* 70
Methods, for communication and telephone apparatus, *99;* 86
Noncurrent-carrying parts, *215C;* 141
Protective, *123;* 95
Purpose, *90;* 67
Scope, *91;* 67
Supply cable termination, *374;* 355
Supply grounding conductors, *239G5;* 261
Supporting structures, *215;* 140
Underground circuits and equipment, *314;* 329

Guarded, *Sec. 2;* 54

Guarding
Live parts, *124;* 96
Where required, *124A;* 96
Shaft end, pulleys, belts, and suddenly moving parts, *122;* 95
Supply conductors, *234C2;* 195

Guards
Location of physical isolation, *124C1;* 96
Shields or enclosures, *124C2;* 97
Strength, *124B;* 96
Types of, *123C;* 96

Guy anchors, *261C;* 293

Guyed poles
Grade D construction, *262A1c, 262A2c;* 307, 308
Grades B and C construction, *261A2a(5), 261A2b(3);* 288, 290

Guying, *264;* 320
Point of attachment, *264C;* 320
Strength, *264B;* 320
That is required but cannot be installed on crossing pole, *262C5;* 309
Where used, *264A;* 320

Guy insulators, *279;* 324
Corrosion protection, *279A3;* 325
Electrical strength, *279A1b;* 324
Installation, *279A2b;* 325
Material, *279A1a;* 324
Mechanical strength, *279A1c;* 324
Properties of, *279A1;* 324
Use of, *279A2;* 325

Guys, *262C;* 308
Fastenings, *264D;* 320
Grade N construction, *263B;* 317
Grounding, *215C2;* 141
Method, *92C2;* 69

Guys *(contd)*
Longitudinal guys, *262C3;* 308
Markers (Guards), *264E;* 320
Overload capacity factors, *Table 262-2;* 315
For guys at angles in line, *Table 262-3;* 315
Side guys, *262C2;* 308
Strength requirements, *261C, 262C4;* 293, 309

H

Handhole, *Sec. 2, 323;* 55, 335
Covers, *323D;* 337
Strength, *323A;* 335

High-voltage contact conductors, *225B;* 151

Highways
Clearance to supporting structures, *231B;* 155
Direct buried cable, *351C4;* 347
Underground conduits, *320A3;* 331

Horizontal clearance. *See* Clearance, horizontal

Horizontal loading component, *251B1;* 280

I

Ice, combined ice and wind loading, *250B;* 276

Identification
Of equipment on supporting structures, *220E;* 147
Of overhead conductors, *220D;* 147
Of supporting structures, *217A3;* 143

Illumination, *111;* 90
Attachment plugs and receptacles for general use, *111D;* 90
Battery areas, *145;* 127
Emergency lighting, *111B;* 90
Fixtures, *111C;* 90
Levels, *Table 111-1;* 91
Receptacles in damp or wet locations, *111E;* 90
Under normal conditions, *111A;* 90

Inch-foot-pound values, *017;* 48

Induced voltages, safety rules, *212;* 139

Installation
Safety rules for overhead electric supply and communication lines, *Part 2;* 138
Surge arresters, *193;* 137
Temporary overhead, *014B;* 47

Instrument transformer, *93C3;* 73

Insulating covering, *124C6;* 97

Insulating jacketed grounded conductor, *354C3;* 351

Insulation, *Sec. 2;* 55
Conductor terminations, *163A;* 130
Shielding, *Sec. 2;* 55

Insulators, *Sec. 2;* 55
Application of rules, *270;* 321
Factory tests, *274;* 323
Grade of construction, *243C;* 275
Insulation level, *273;* 322
Requirements for, *Table 273-1;* 322
Material and marking, *271;* 321
Mechanical strength of, *277;* 323
Protection against arcing and other damage, *276;* 323
Ratio of flashover to puncture voltage, *272;* 322
Special applications, *275;* 323
For constant-current circuits, *275A;* 323
For single-phase circuits, *275B;* 323

Intent, *015;* 47

Introduction to NESC, *Sec. 1;* 45

Isolated, *Sec. 2;* 55
By elevation, *Sec. 2;* 55

Isolated phase bus, *181B;* 136

Isolating switch, *Sec. 2;* 53

Isolation, non-shielded insulated conductors, *162;* 130

Isolator, *Sec. 2;* 55

J

Jacket, *Sec. 2, 331;* 55, 340

Jacketed grounded conductor, *354C2;* 350

Jacketed multiple-conductor cables, *239G2;* 260

Joint use, *Sec. 2;* 55
- Manholes, *341B3b;* 343
  - Clearance between supply and communications facilities in, *Table 341-1;* 344
- Poles, climbing, *432;* 371
- Structures, *222;* 148
  - Longitudinal loading at crossings, *252C2;* 284

L

Ladders, *323F;* 337
- Space, *234C5;* 197

Lateral communication conductors, *239F;* 259

Lateral conductors
- Clearance, *Table 239-1;* 264, 265
- Conductors not in conduit, *239C;* 257

Lateral supply conductors
- Clearance from other facilities and surfaces on same support, *239;* 257
- Of less than 300 volts, *239E2c;* 258
- Location relative to climbing spaces, working spaces, and pole steps, *239B;* 257
- On supply line structures or within supply space on jointly used structures, *239E;* 258

Latticed structures, *252B2c;* 282

Lighting
- Decorative, *217A5;* 143
- Streets and areas, *263H;* 318
  - Vertical clearance, *232B4;* 157

Line insulation, *Sec. 27;* 321

Lines, *Sec. 2;* 55
- Communication, *Sec. 2;* 55
- Electric supply, *Sec. 2;* 56
- Inspection and tests of, *214;* 139
  - When in service, *214A;* 139
  - When out of service, *214B;* 140

Line supports
- Grades of construction, *243E;* 274
- Loads upon, *252;* 281

Liquid cell batteries, *420G;* 364

Liquid hydrogen systems, *127H;* 114

Liquified petroleum gas (LPG), *127K;* 115
- Storage (Classified areas), *Table 127-4;* 122

Live-line work, *446;* 388
- Bonding and shielding for bare-hand method, *446C;* 389
- Equipment, *446B;* 389
- Training, *446A;* 389

Live parts, clearance from, *Table 124-1;* 100

Loading
- Assumed longitudinal loading, *252C;* 283
- Conductor loading, *251;* 280
  - General, *251A;* 280
  - Loading components, *251B;* 280
- Flammable liquids, *127D;* 110
- Grades B, C, and D, *Sec. 25;* 276
  - General, *250A;* 276
  - Ice and wind loading, *250B;* 276
- Horizontal wind pressures on cylindrical surfaces, *Table 250-2;* 279
- Ice, wind and temperature, *Table 250-1;* 279

Loading *(contd)*
Loads upon line supports, *252;* 281
Maps
Basic wind speed, *Fig 250-2;* 278
Loading overhead lines, *Fig 250-1;* 277
Simultaneous application of loads, *252D;* 285
Temperature and constants, *Table 251-1;* 281

Locking (Of switches), *216C;* 142

Longitudinal guys, *262C3;* 308

Longitudinal strength, *261A2a(3);* 287
Crossarms, *261D4a;* 294
Pin type construction, *261F1;* 295

Longitudinal-strength requirements, *261A4;* 291

Low voltage protection, *Sec. 2;* 56

Luminaries, conductors to, *239E2b;* 258

M

Maintenance of overhead electric supply and communication lines, safety rules, *Part 2;* 138
Maintenance replacements, *013B2;* 46

Manholes, *Sec. 2, 323;* 56, 335
Access, *323C;* 336
Cables in, *341B;* 342
Clearance, *341B2;* 342
Covers, *Sec. 2, 323D;* 56, 337
Dimensions, *323B;* 336
Drainage, *323G;* 338
Grating, *Sec. 2;* 56
Guarding, *423A;* 369
Identification, *323J, 341B3;* 338, 343
Ladder requirements, *323F;* 337
Mechanical protection, *323I;* 338
Strength, *323A;* 335
Supports, *341B1;* 343
Testing for gas in, *423B;* 369
Underground conduits in vicinity of, *322B6;* 335
Ventilation, *323H;* 338

Mats (Insulating), *124C4;* 97

Mechanical injury, *217A1a;* 142

Mechanical protection
Conductors, *161;* 130
Manholes and vaults, *323I;* 338

Messengers, *262I;* 312
Conductors supported by, *235;* 220
Grounding, *92C;* 69
Sags and tensions, *262I2;* 312
Strength, *262I1;* 312

Metal crossarms, strength requirements, *261D2;* 294

Metal enclosed bus, *181;* 136
General requirements, *181A;* 136
Isolated phase bus, *181B;* 136

Metal enclosed power sitchgear, *180B;* 133
Location, *180Bl;* 133
Space requirements, *180B3, 180B4, 180B5;* 134
Switchgear rooms, *180B2;* 134

Metal-sheathed communication cables, *239H1;* 262

Metal-sheathed cable, insulation for, *163B;* 130

Metal structures
Grade B and C construction, *261A1;* 286
Overload capacity factors, *Table 261-2;* 302
Strength requirements, *261C1;* 293

Metric system (SI), *017;* 48

Mobile hydrogen equipment, *129;* 125

Motor control, *131;* 126

Motors
Adjustable-speed, *130E;* 126
Control centers, *180D;* 135
Speed limit, *130C;* 125

Multiconductor wires, *235;* 220

Multigrounded ac systems, *93C2;* 72

Multi-grounded systems
Ground resistance, *96A3;* 84
Separation of grounding conductors, *97D1;* 85

Multiple messengers, grounding, *215C3;* 141

N

Natural gas (Methane), *127L;* 115
Classified areas (Equipment), *Table 127-5;* 124

Neutral conductors
Clearance, *230E;* 154
Grade of construction, *242E;* 268
Grounding, *314C1;* 329

Noncurrent-carrying parts
General, *215C1;* 141
Grounding, *215C;* 141
Guys, *215C2;* 141
Multiple messengers, *215C3;* 141

Non-magnetic conduits, *181B2;* 136

Notes, *015D;* 47

O

Obstructions, of supporting systems, *217A4;* 143

Oil—circuit breakers, reclosers, and switches containing, *172;* 131

Open supply conductors, *014A1a;* 47

Open vertical conductors, clearance between pole center and, *Table 239-2;* 264, 265

Open-wire communication conductors, strength requirements, *261J;* 299

Operating rods, *239I;* 263

Overhead conductors, *220D;* 147

Overhead electric supply, safety rules for installation and maintenance, *Part 2;* 138

Overhead line operating procedures, *422;* 368
Checking structures before climbing, *422B;* 368
Installing and removing wires or cables, *422C;* 368
Setting, moving, or removing poles in or near energized electric supply lines, *422A;* 368

Overhead supply lines, and equipment approach distances to exposed energized parts, *Table 431-1;* 371

Overhead temporary installations, *014B;* 47

Overload capacity factors
For Foundations and settings, *Table 261-4;* 305
For guys, *Table 262-2;* 315
At angles in line, *Table 262-3,* 315
For metal and prestressed concrete structures, *Table 261-2;* 302
For reinforced concrete structures, *Table 261-1;* 301
Unguyed wood poles, *Table 262-1A;* 313
When installed, *Table 262-1B;* 314
For wood structures, *Table 261-3A;* 303
When installed, *Table 261-3B;* 304

Oxygen (Bulk installations), *127J;* 114

P

Pad-mounted equipment, *Sec. 2, 381G;* 56, 356

Paired conductors
- Communication (grades B and C), *261L;* 299
- Not supported on messenger, *261L2;* 299
- Supported on messenger, *261L2;* 299
- Without messengers, *262H2;* 311

Pin construction, strength requirements, *261F;* 295

Pins, *262F;* 310
- Grade of construction, *243C;* 275
- Size, *262F2;* 310
- Strength, *262F1;* 310

Piping systems, *95B;* 82

Plowing, *353B;* 348

Pole center, clearance between open vertical conductors and, *Table 239-2;* 264, 265

Pole risers, *362;* 353

Poles
- Average strength of, *261A2a(7), 261A2b(5);* 288, 290
- Butt plates and wire wrap (Grounding electrode), *94B4;* 79
- Grade D construction, *262A;* 307
- Grade N construction, *263A;* 317
- Guyed poles, *261A2a(5), 261A2b(3);* 288, 290
- Spliced and stub-reinforced, *261A2a(6), 261A2b(4), 261A1d;* 288, 290, 307
- Vertical conductor clearances, *239E2d;* 259

Pole steps, location of conductors relative to, *239B;* 257

Prestressed concrete structures, *Sec. 2;* 56
- Overload capacity factors, *Table 261-2;* 302
- Strength requirements, *261C1;* 293
  - Grades B and C construction, *261A1;* 286

Primary surge arresters, *93C4;* 73

Prime movers, *130A;* 125

Protection
- Communication protective requirements, *315;* 329
  - Means of protection, *315B;* 330
  - Where required, *315A;* 329
- Control circuits, *130F;* 126
- Current-transformer secondary circuits, *150;* 128
- Electrical, *160;* 129
- Grounding conductor, *93D;* 75
- Mechanical, *161;* 130
  - Near ground, *239D;* 257
- Short circuits, *133;* 126
- Supporting structures, *217A1;* 142

Protective
- Communications apparatus, *223;* 148
  - Means of protection, *223B;* 148
  - Where required, *223A;* 148
- Devices and equipment, *411B;* 361
- Grounding, *123;* 95
  - Method, *123;* 95
  - Provision for grounding during maintenance, *123;* 95
  - Supply station equipment, *123;* 95
- Grounds, *445;* 387
  - Installing, *445A;* 387
  - Removing, *445B;* 388
- Identification of supply circuits, *411E;* 362
- Inspection and testing of devices, *411C;* 361
- Methods, *411A;* 360
- Warning signs, *411D;* 362

Pulling
- Iron, *Sec. 2;* 57
- Tension, *Sec. 2;* 57

Q

Qualified, *Sec. 2;* 57

R

Raceway, *Sec. 2;* 57

Racks
Storage batteries, *143;* 127
Strength requirements, *262E;* 310

Rail cars
Clearance to wires, conductors and cables to, *234I;* 202
Rail car clearance, *Fig 234-3;* 206

Railings, guards, *124C3, Fig 124-2;* 97, 99

Railroad crossings, prevention of loss of contact at grade, *225D;* 151

Railroad tracks
Clearance from supporting structures, *231C;* 156
Direct buried cable, *351C3;* 346
Grades of construction for communication conductors crossing over, *Table 241-1;* 267
Underground conduits crossing, *320A5;* 331

Railway crossings, *225;* 151

Railway feeder circuit conductors, *242B;* 268

Railways, vertical clearance, *232;* 156

Random separation, *Sec. 2, 354;* 57, 349

Readily climbable, *Sec. 2;* 57

Reclosers, *Sec. 17;* 131
Application, *171;* 131
Arrangement, *170;* 131
Oil in, *172;* 131

Recommendation, *015C;* 47

References, *Sec. 3;* 63

Regulators, *Sec. 15;* 128
Indoor installation, *152B;* 129
Location and arrangement, *152;* 128
Outdoor installations, *152A;* 128

Reinforced concrete structures
Overload capacity factors, *Table 261-1;* 301
Strength requirements, *261A1, 261C2;* 286, 293

Relative circuits, *220B2;* 144

Relative levels (Of conductors), *220;* 144
Standardization of levels, *220A;* 144
Supply and communication conductors, *220B;* 144
Preferred levels, *220Bl;* 144
Supply lines of different voltage classifications, *220C;* 146
At crossings or conflicts, *220Cl;* 146
On structures used only by supply conductors, *220C2;* 147

Remotely operable, *Sec. 2;* 57

Ridge-pin conductors, climbing space near, *236I;* 247

Risers, *Sec. 36;* 352
General, *360A;* 352
Installation, *361;* 353
Pad-mounted installations, *363;* 353
Pole risers, *362;* 353

Roadways, *Sec. 2;* 57
Clearance to supporting structures, *231B;* 155
Vehicle load, *Fig 323-1;* 339
Vertical clearance, *232;* 156

Rotating equipment, *Sec. 13;* 125

Rule 230C1, *239G1;* 260

Rural districts, *Sec. 2;* 57

S

Sag, *Sec. 2;* 57, 58
  Apparent, *Sec. 2;* 58, 59
  At any point in span, *Sec. 2;* 59
  Conductors of different sags on same support, *235C2b;* 225
  Conductor tension, *262H3;* 311
  Final, *Sec. 2;* 58
    Unloaded, *Sec. 2;* 58
  Initial unloaded, *Sec. 2;* 58
  Line conductors, *234H1;* 201
  Maximum total, *Sec. 2;* 58
  Messengers, *262I2;* 312
  Open supply conductors, *261H2;* 297
  Total, *Sec. 2;* 58

Separation, *Sec. 2;* 59
  Grounding conductors, circuits, *97;* 84

Service drop, *Sec. 2;* 59
  Grade N construction, *263E;* 317
    Cabled service drops, *263E3;* 318
    Size, *263E1, 263E3a;* 317, 318
    Tension, *263E3b;* 318
  Of 750 V or less, sizes of, *Table 263-2;* 319

Settings
  Overload capacity factors, *Table 261-4;* 305
  Strength requirements (Grades B and C construction), *261B;* 292

Sewers, *320B3;* 332

Shall, *015A;* 47

Sheaths, *331;* 340
  Bonding, *93E6;* 76

Shielded cable, *163B;* 130

Shielding, *332;* 340
  General, *332A;* 340
  Material, *332B;* 341

Shields (Guards), *124C2;* 97

Short circuit protection, *133;* 126

Should, *015B;* 47

Shoulder (Roadway), *Sec. 2;* 59

Sidearm construction, *239E2a;* 258

Side guys, *262C2;* 308

Side-wall pressure, *Sec. 2;* 59

Single-grounded systems
  Grounding conductor, *93C1;* 72
  Ground resistance, *96A2;* 84
  Separation, *97D1;* 85

Spacer cable, *Sec. 2;* 50

Spacings
  Maintenance, *230;* 155
  Measurement, *230B;* 153

Span length, *Sec. 2;* 59
  Transverse load, *252B4;* 283

Span wire, *Sec. 2;* 59
  Clearance, *238C;* 254
  Conductors supported by, *235;* 220
  Vertical clearance from communication lines, *Table 238-2;* 256

Span-wire insulators, *279B;* 325
  Insulation level, *279B1b;* 325
  Location, *279B2b;* 326
  Material, *279B1a;* 325
  Mechanical strength, *279B1c;* 325
  Properties of, *279B1;* 325
  Use of, *279B2;* 326

Speed control devices, *130;* 125

Speed limit for motors, *130C;* 125

Splices
  Conductors, *262H4;* 311
  Open supply conductors, *261H3;* 298

Standards, references, *Sec. 3;* 63

Steam lines, *320B6;* 333

Stopping devices, *130;* 125
  Automated overspeed trip device, *130A;* 125
  Manual, *130B;* 125

Storage areas, for flammable liquids, *127C;* 110

Storage batteries, *Sec. 14;* 127
Floors in battery areas, *144;* 127
General, *140;* 127
Illumination, *145;* 127
Location, *141;* 127
Racks, *143;* 127
Service facilities, *146;* 128
Ventilation, *142;* 127

Streets
Clearance to supporting structures, *231B;* 155
Direct buried cable, *351C4;* 347
Lighting
Equipment, *263H;* 318
Vertical clearance, *232B4;* 157
Underground conduits, *320A3;* 331

Strength
Foundations and settings, *261B;* 292
Grades B and C construction, *261;* 286
Grounding conductor, *93C, 93C9;* 72, 74
Guying and bracing, *264B;* 320
Longitudinal-strength requirements for sections of high grade in lines of lower grade construction, *261A4;* 291
Preliminary assumptions, *260;* 285
Requirements, *Sec. 26;* 285
Transverse requirements where side guying is required, *261A3;* 291

Stringing loads, *252C5;* 284

Structure loading, *252B2;* 282

Structures
Conflict, *Sec. 2;* 59
Joint use of, *222;* 148

Submarine crossings, *320A6;* 332
Direct buried cable, *351C5;* 347

Sulfur dust locations, *127I;* 114

Supply cable, *Sec. 33;* 340
Clearance, *230C;* 153
Emergency installations, *014A1b;* 47
Fiber optic, *Sec. 2;* 54
Grades of construction, *241A;* 266
Random separation, *354A;* 349
In direct burial, *354C;* 350

Supply cable messengers, *261I;* 298

Supply cable terminations, *Sec. 37;* 353
Clearances in enclosures or vaults, *373;* 354
General, *370A;* 353
Grounding, *374;* 355
Identification, *372;* 354
Support at terminations, *371;* 354

Supply circuits, *242E;* 268

Supply conductors
Attached to buildings or other installations, *234C3;* 196
Grades of construction for supply conductors alone, at crossing, or on same structures with other conductors, *Table 242-1;* 270
Guarding, *234C2;* 195
Open, *014A1a;* 47
Size, *261H1;* 297

Supply employee rules, *Sec. 44;* 372
Approach to energized conductors or parts, *441;* 372

Supply equipment, *Sec. 2;* 59

Supply line conductors, *263D, Table 263-1;* 317, 319

Supply lines, grades of construction for communication conductors crossing over, *Table 241-1;* 267

Supply station, *Sec. 2;* 59

Supply support arm, vertical supply conductors suspended from, *239G4;* 260

Support arms, grade of construction for, *243B;* 274

Supporting structure, *Sec. 2, 217A;* 60, 142
Clearance from other objects, *231;* 155
Grades B and C construction, *261;* 286
Grounding, *215;* 140
Identification of equipment on, *220E;* 147
Protection, *217A1;* 142
Unusual conductor supports, *217B;* 143
Vertical and lateral conductor clearances from, *239F2;* 259

Surge arresters, *Sec. 19;* 136
General requirements, *190;* 136
Grounding, *215B4, 314C3;* 140, 329
Grounding conductors, *93C4, 192;* 73, 137
Grounds, *92B2b;* 68
Indoor locations, *191;* 136
Installations, *193;* 137

Surge protection wires, *242F;* 269

Susceptiveness, *Sec. 2;* 60

Suspension insulators
Clearances, *235E2;* 227
Horizontal clearance, *235B2;* 222

Swimming pools
Clearance, *234E1;* 198
Direct buried cable, *351C1;* 346

Switchboards, *Sec. 2;* 60
Control, *180E;* 135
Dead front power, *180C;* 135

Switches, *Sec. 2, Sec. 17, 173;* 60, 131, 132
Application, *171;* 131
Arrangement, *170, 216;* 131, 142
Capacity, *173A;* 132
Disconnecting, *173B;* 132
Oil in, *172;* 131
Visible break switches, *173C;* 132

Switchgear, *Sec. 18;* 132
Switchgear assemblies, *180;* 132
General requirements, *180A;* 132
Metal enclosed power switchgear, *180B;* 133

Switching control procedures, *442;* 380
Designated person, *442A;* 380
Operations at stations, *442C;* 380
Re-energizing after work, *442D;* 380
Repeating oral messages, *442G;* 381
Restoration of service after automatic trip, *442F;* 381
Specific work, *442B;* 380
Tagging electric supply circuits, *442E;* 381

T

Tag, *Sec. 2;* 60

Taps
Conductors, *262H4;* 311
Open supply conductors, *261H3;* 298

Telephone, grounding methods for, *99;* 86

Temporary overhead installations, *014B;* 47

Tensions
Messengers, *262I2;* 312
Open supply conductors, *261H2;* 297
Unloaded
Final, *Sec. 2;* 60
Initial, *Sec. 2;* 60

Termination, *Sec. 2;* 60
Conductor, *163;* 130

Thermal protection, *352D;* 348

Third rail, *225C;* 151

Through bolts
Communication conductor, *239H4;* 263
Vertical conductor clearance from, *239G6;* 261

Tools, employee rules for, *420H;* 364

Traffic signal brackets, *238D;* 254

Transformers, *Sec. 15;* 128
Instrument, secondary circuits of, *151;* 128
Location and arrangement, *152;* 128
Indoor installation, *152B;* 129
Outdoor installation, *152A;* 128
Secondary circuits exceeding 600 V, *150;* 128
Vault, *Sec. 2;* 60

Transverse
Loading, from conductors and messengers, *252B1;* 282
Strength, *261A2a(2);* 287
Where side guying is required, *261A3;* 291

Transversely weak section, in construction, *261F1d;* 296

Traveled way, *Sec. 2;* 60

Tree trimming, *218;* 144

Trenching, *353A;* 348

Trolley-contact conductor
Fastenings, *225A;* 151
Grade N construction, *263F;* 318
Grade of construction for, *242B;* 268
Guards under bridges, *234D2;* 198
Open supply conductors, *261H3;* 298
Paired communication conductors above, *261L2b;* 300

Tunnels, *Sec. 39;* 357
Environment, *391;* 358
General, *390;* 357
Multiple use, *391B;* 358
Underground conduits, *320A4;* 331

U

Underground conduit system, *Sec. 32;* 330
Bridges and tunnels, *320A4;* 331
Crossing railroad tracks, *320A5;* 331
Ducts and joints, *322;* 333
General, *322A;* 333
Installation, *322B;* 334
Excavation and backfill, *321;* 333
General, *320A1;* 331
Highways and streets, *320A3;* 331
Location, *320;* 331
Natural hazards, *320A2;* 331
Routing, *320A;* 331
Separation from other underground installations, *320B;* 332

Underground grounding conductor, *93E;* 75

Underground line operating procedures, *423;* 369
Excavation, *423D;* 370
Flames, *423C;* 369
Guarding manhole and street openings, *423A;* 369
Identification, *423E;* 370
Operation of power driven equipment, *423F;* 370
Testing for gas in manholes and unventilated vaults, *423B;* 369

Underground lines, *Sec. 31;* 327
Accessibility, *312;* 328
Induced voltage, *316;* 330
Inspections and tests of, *313;* 328
When in service, *313A;* 328
When out of service, *313B;* 328
Installation and maintenance, *311;* 327
Purpose of code, *300;* 327
Referenced sections, *321O;* 327
Scope of code, *301;* 327

Underground structures, equipment in, *382;* 356

Ungrounded
Separation of grounding conductors, *97D1;* 85
Supply systems, *354C4;* 351

Uniform position—switches, *216D;* 142

Units of measure, *017;* 48

Unguyed poles
Foundations and settings, *262B;* 308
Overload capacity factors, *Table 262-1A;* 313
And strength factors when installed, *Table 262-1B;* 314
Strength (Grade D construction), *262A1b, 262A2b;* 307

Urban districts, *Sec. 2;* 61

Utility, *Sec. 2;* 61

Utility interactive system, *Sec. 2;* 61

Utility tunnel, *323E;* 337

Utilization equipment, *Sec. 2;* 61

V

Vaults, *Sec. 2, 323;* 61, 335
Access, *323E;* 337
Cables in, *341B;* 342
Clearance between supply and communications facilities in, *Table 341-1;* 344
Drainage, *323G;* 338
Electrical clearances in, *373;* 354
Ladder requirements, *323F;* 337
Strength, *323A;* 335
Ventilation, *323H;* 338

Ventilation
Manholes and vaults, *323H;* 338
Storage batteries, *142;* 127

Vertical clearance. *See* Clearance, vertical

Vertical communication conductors
On communication line structures or within communication space on jointly used structures, *239F;* 259
Passing through supply space on jointly used structures, *239H;* 262

Vertical conductors
Clearance, *Table 239-1;* 264, 265
Climbing space past, *236H;* 247

Vertical loading component, *251B1;* 280

Vertical strength, *261A2a(2);* 287

Vertical supply conductors
Clearance from other facilities and surfaces on same support, *239;* 237
Conductors not in conduit, *239C;* 257
Of less than 300 volts, *239E2c;* 258
Location relative to climbing spaces, working spaces, and pole steps, *239B;* 257
Mechanical protection near ground, *239D;* 257
Passing through communication space on jointly used line structures, *239G;* 260
Sidearm construction, *239E2a;* 258
On supply line structures or within supply space on jointly used structures, *239E;* 258

Visible break switch, *173C;* 132

Voltage, *Sec. 2;* 61
Circuit
Constant current circuit, *Sec. 2;* 61
Effectively grounded circuit, *Sec. 2;* 62
Not effectively grounded, *Sec. 2;* 61
To ground, *Sec. 2;* 62
Ground circuit, *Sec. 2;* 62
Ungrounded circuit, *Sec. 2;* 62

W

Waiver, *014;* 46

Wall plates (Isolated-phase bus), *181B4;* 136

Water
Beaches and waterways restricted to swimming—clearances, *234E;* 199

Water *(contd)*
Skiing areas—clearance to wires, conductors, cables, *234E3;* 199
Underground conduits for water lines, *320B4;* 333

Wheel load area, *Fig 323-2;* 339

Wind
Combined ice and wind loading, *250B;* 276
Extreme wind loading, *250C;* 276

Wire
Sizes with respect to loading district and span length, *Table 262-5;* 316
Vertical and lateral conductor clearance from, *239F1;* 259
Vertical clearance, *232;* 156

Wire gages, *Sec. 2;* 62

Wood
Crossarms, *261D;* 293
Permitted stress level, *261A2b(1);* 289
Poles
Grade D construction, *262A;* 307
Strength requirements, *261C2;* 293
Structures
Overload capacity factors for, *Table 261-3A;* 303
And strength factors when installed, *Table 261-3B;* 304
Strength, *261A2, 261A2b(2);* 286, 289

Working space, *Table 125-1, 237;* 107, 252
Dimensions of, *237B;* 252
Electrical equipment, *125;* 106
Guarding energized equipment, *237E;* 253
Location, *237A;* 252
Location of buckarms relative to, *237D;* 252
Location of vertical and lateral conductors relative to, *237C, 239*B; 252, 257
Obstruction by buckarm, *Fig 237-1;* 254
Working clearances from energized equipment, *237F;* 253

Procedure for Revising the National Electrical Safety Code

1. Preparation of Proposals for Amendment
   - 1.1 A proposal may be prepared by any:
     - a. substantially interested person
     - b. interested organization
     - c. NESC Subcommittee
     - d. member of the NESC Committee or its subcommittees
   - 1.2 Proposals shall be submitted to: Secretary, National Electrical Safety Code Committee (at the address listed in the time schedule for revision). Copies must be suitable for reproduction and shall be on forms for new proposals available from the secretariat.
   - 1.3 Each separate topic shall begin on a separate form, printed or typed on one side only.
   - 1.4 The proposal shall consist of the following:
     - a. a statement, in NESC rule form, of the exact change, rewording, or new material proposed
     - b. the name of the submitter (organization or individual as applicable)
     - c. supporting comment, giving the reasons why the NESC should be so revised (see SAMPLE submittal)

2. The NESC Secretariat shall:
   - a. Acknowledge receipt of proposals for revision. (If the submitter does not receive an acknowledgment within 30 days of mailing his proposal, the secretariat should be contacted.)
   - b. Distribute to each member of the appropriate NESC Subcommittee all of the proposals received, arranged in a coordinated sequence.

3. Subcommittee Recommendation

   The NESC Subcommittee responsible shall consider each proposal and take one or more of the following steps:
   - a. Endorse the proposal as received.
   - b. Prepare a proposed revision or addition for the NESC (this may be a coordination of several comments, or a committee consensus on a modification of a proposal).
   - c. Refer the proposal to a technical working group for detailed consideration.
   - d. Request coordination with other NESC Subcommittees.
   - e. Recommend rejection of the proposal for stated reasons.

   For each item, a subcommittee voting statement shall be prepared, accompanied by all members' statements concerning their votes (cogent reasons are required for negative votes). Steps (c) and (d) are intended to result, eventually, in a proposal of category (b).

   Action under steps (c) or (d) shall be completed and reported to the subcommittee before the end of the 5-month public review period, if the item is to be included in the upcoming revision.

4. Preprint of Proposals

   The NESC Secretariat shall organize and publish a preprint of proposed C2 revisions including:
   - a. The original proposal as received from the submitter
   - b. The recommendation of the subcommittee with respect to the proposal (including a voting statement and subcommittee members' statements)
   - c. Copies of submittal forms for comments

   The preprint shall be distributed to all members of NESC Subcommittees and the representatives of the organizations comprising the NESC Committee. Copies shall be available for sale to other interested parties. Notice of availability of the preprint shall be submitted to ANSI for publication in ANSI Standards Action. The preprint shall carry information on how to submit comments on the proposals and the final date for such submissions.

5. Final Processing of Proposed Revisions and Comments
   - 5.1 Following the public review period, the secretariat shall organize and distribute for subcommittee consideration all comments received.
   - 5.2 The preprint and the comments received shall be reconsidered by the subcommittees. No new change proposals may be considered. Substantive changes in the ballot draft from the recommendations in the preprint shall be appropriately noted.
     - a. The subcommittee may recommend adoption or rejection of the proposal by majority vote;
     - b. When extended technical consideration or resolution of differing or conflicting points of view is necessary, the subcommittee shall refer the problem to a working group of the subcommittee for proposed resolution. If expeditious consideration is not possible, the subject shall be held on the docket.

     Each working group shall provide, to its parent subcommittee, recommendations on matters considered as a result of subcommittee referrals under items 5.2(a) and 5.2(b).

     Each subcommittee shall prepare a report showing its proposed revisions and all items to be held on the docket together with a plan for their disposition.
   - 5.3 The Secretary shall provide commentors with copies of reports of action taken on the rules affected by their comments, and shall make all such reports available for examination upon request.

6. Final Approval
   - 6.1 Based upon the subcommittee reports, the NESC Secretariat shall prepare a draft revision of the NESC (C2) and distribute copies to the:
     - a. NESC Committee for approval by a 6-week letter ballot
     - b. American National Standards Institute, Board of Standards Review, for concurrent 60-day public review
   - 6.2 Comments received in response to the letter ballot and public review shall be referred to the executive subcommittee for resolution or referral to the appropriate subcommittee. Those items on which consensus cannot be reached shall be referred to the appropriate subcommittee for consideration during the next revision cycle. Unless a consensus for revision is established, the requirements of the current edition shall carry over to the proposed edition.
   - 6.3 *ANSI Approval.* When resolution of comments received in response to letter ballot and public review has been completed, the proposed new edition of the NESC shall be submitted to ANSI for approval.

Request for Change
National Electrical Safety Code

Rule 233A1b(5), revise to read:

(5) Where crossing is not at midspan of the upper conductor ~~and under conditions where the upper span exceeds those specified in the Rule 233A1b(2)~~, the additional sag determined in accordance with Rules 233A1b(3) and 233A1b(4) may be reduced by multiplying ~~the additional sag determined by Rules 233A1b(3) and 233A1b(4)~~ by the following factors.

Supporting Comment: It is believed that the suggested revision clarifies the intent of the Rule. Additionally, the revision properly addresses Rule 233A1b(4), which applies to upper conductors designed to operate at a conductor temperature above 120 °F (50 °C), *regardless of span length* [based on companion proposal for Rule 233A1b(4)].

Time Schedule for the Next Revision National Electrical Safety Code

Sept 30, 1990	Final date for receipt of proposals from the public for revision of the 1990 edition of the National Electrical Safety Code, preparatory to the publication of a 1993 edition. Proposals should be forwarded in the prescribed form to: Secretary National Electrical Safety Code Committee IEEE Standards Office 445 Hoes Lane, P.O. Box 1331 Piscataway, NJ 08855-1331
Oct 29 to Nov 23, 1990	NESC Subcommittees consider proposals for Code changes and prepare their recommendations.
Apr 15, 1991	Preprint of Proposed Amendments for incorporation into the 1993 edition of the National Electrical Safety Code published for distribution to the NESC Committee and other interested parties.
Apr 15 to Sept 30, 1991	Period for study of proposed amendments and submittal by interested parties of recommendations concerning the proposed amendments. Submit recommendations to the Secretary, NESC Committee, at the above address.
Oct 28 to Nov 22, 1991	Period for NESC Subcommittee Working Groups and NESC Subcommitees to reconsider all recommendations concerning the proposed amendments and prepare final report.
* Feb 1, 1992	Proposed revision of NESC, ANSl C2, submitted to NESC Committee for letter ballot and to the American National Standards Institute for concurrent public review.
May 15, 1992	NESC Committee approved revisions of NESC submitted to American National Standards Institute for approval as an ANSl standard.
Aug 1, 1992	Publication of the 1993 edition of the National Electrical Safety Code.

* Since the document to be submitted to the NESC Committee for letter ballot will be a complete draft of the 1993 edition subject to editorial changes only, it can serve as the starting point for the next round of revisions, that is, those proposed for incorporation into the 1996 edition. Copies will be available for purchase from the secretary at the above address.

Corrections to Third Printing

p. 47 ..... In Rule 014A1c, the spelling of the word "shall" was corrected. In Rule 015B, the spelling of the word "conditions" was corrected.

p. 108 .... In Rules 127A1 and 127A2, "105 ohm-cm" was corrected to "10^5 ohm-cm."

p. 109 .... In Rule 127A5, "105 ohm-cm" was corrected to "10^5 ohm-cm."

p. 134 .... In Rules 180B5 and 180B6, "181B4" was corrected to "180B4."

p. 189 .... In Table 233-1 M, row 3, the fifth column, "2.0" was corrected to read "0.60;" in row 4, all values of "2.40" were corrected to read "1.80."

p. 196 .... The word "to" was removed from Rule 234C3.

p. 277 .... The equations should read as follows:

$$\text{pressure in lb/ft}^2 = 0.00256\,(v_{mi/h})^2$$

$$\text{pressure in pascals} = 0.613\,(v_{m/s})^2$$

Figure 250-1 had inadvertantly disappeared from the Second Printing, and was replaced on the page.

p. 289 .... In Rule 261A2b(2) EXCEPTION 2, the word "length" was corrected to the word "strength."

p. 373 .... In Rule 441A2, "100 feet (300 meters)" was corrected to read "1000 feet (300 meters)."

Tentative Interim Amendment 90-1
to the
National Electrical Safety Code
ANSI C2-1990

In accordance with Section 13 of its Procedures, the National Electrical Safety Code Committee has issued the following Tentative Interim Amendment (TIA) to ANSI C2, National Electrical Safety Code, 1990 Edition. The TIA was issued by the Secretariat on June 7, 1990, as a result of a proposal submitted by a member of Technical Subcommittee 4 on Overhead Line Clearances.

A Tentative Interim Amendment is tentative because it has not been processed through the entire standards-making procedure. It is interim because it is effective only between editions of the code. A TIA automatically becomes a Proposal of the proponent for the next edition of the code; as such, it is then subject to all the procedures of the standards-making process.

1. In the third line of the definition of "*V*" in Rule 234I, change "22 ft (6.7 m)" to "20 ft (6.1 m)."
2. In footnotes (2) and (16) to Table 232-1 FT, change "22 ft" to "20 ft" (3 locations). In the metric table (Table 232-1 M), footnotes (2) and (16), change "6.7 m" to "6.1 m."